Homestead Tsunami

Homestead Tsunami

Good for Country, Critters, and Kids

By Joel Salatin

Foreword by Amy K. Fewell

Polyface, Inc.
Swoope, Virginia

This publication is designed to provide accurate and authoritative information in regard to the subject matter covered. It is sold with the understanding that the publisher is not engaged in rendering legal, accounting or other professional service. If legal advice or other expert assistance is required, the services of a competent professional person should be sought.

From a declaration of principles jointly adopted by a committee of the American Bar Association and a committee of publishers.

Homestead Tsunami: Good for Country, Critters and Kids

First Edition, 2023
© 2023 Joel Salatin
All rights reserved.
ISBN: 978-1-7336866-3-1
Library of Congress Control Number: 2023911437

Edited by Jamie Shaver Marsh.
Final edits by Jennifer L. Dehoff and Ashley N. McLain.
Cover photos by Rachel Salatin.
Cover design & book layout by Jennifer Dehoff Design.
All rights reserved.

Printed and bound in the USA.

I dedicate this book to my paternal grandfather,
Frederick Salatin,
known affectionately by friends as "Fritz,"
whose large garden, compost, chickens, and mobile sprinklers
encouraged me to love visceral backyard abundance.

He was working in his garden one morning, felt tired,
and decided to take a nap under the shade of a tree,
and never woke up. What a way to go.

JOEL SALATIN and his family operate Polyface Farm in Virginia's Shenandoah Valley. His parents purchased the farm in 1961 and developed the basic principles of design and production that now show 60 years' refinement.

Joel gravitated toward communication activities in high school and college, graduating with a BA degree in English, and after a brief journalism hiatus returned to the family farm full time Sept. 24, 1982.

Editor of *The Stockman Grass Farmer* magazine, he writes and speaks around the world on food and farm issues. With a long track record of innovation and excellence, Polyface holds educational seminars, farm tours, day camps and events to encourage duplication and understanding.

This is Joel's 16th book.

CONTENTS

Foreword

By Amy K. Fewell

As I watch the sun rising over the field—the dew delicately dripping down the blades of grass—I can't help but take a deep breath of solitude. There is something special about taking in a deep breath of clean, crisp, morning air on the homestead. The turkeys and chickens peck the ground, looking for their next treat. The cows quietly graze, and if you're close enough you can hear the faint ripping of the dew-covered grass from the ground beneath them. Solitude. Happiness. Abundance. Blessing. These are the words I think of on quiet mornings like these.

Of course, this idyllic imagery of our homestead is not always true or realistic. Most mornings there are children impatiently waiting for breakfast, a sassy heifer bossing me around, and food from the garden staring me down ready to be preserved—except, I'm not ready. I'm busy rushing around keeping children fed and clothed, answering the phone, scrolling through my newsfeed for work, or wondering where I put whatever was in my hands five seconds before. Don't be fooled, there is a beautiful side to homesteading, but there is also a beautifully balanced reality.

It's in those quiet moments though—be it in the garden or washing dishes—that I find the best of solitude. The place where my mind wanders and I truly get a glimpse of why we do what we do. A place where the created collides with the Creator. The thought of "why" we homestead. Why we break a sweat over and over again to be producers instead of consumers. Why we endure a lack of modern amenities like the grocery store and pharmacy, and try to live a healthier way of life through natural remedies and farming. Why we break bread with others on our homestead, even during the busiest seasons of life.

Our "why" began in 2010, when our oldest son was diagnosed with childhood asthma. I saw the amount of medicine they wanted to put him on, and something inside of me knew, this just wasn't a way to live. Our "why" grew over the years, even after our son was healed of asthma. It grew into wanting better, healthier food. It grew into food security and financial security. It grew into a better, simple way of living. It grew into creating community. It grew into a conviction of a holier way to live.

My guess is, if you've picked up this book to read, you either already know your "why", or you're about to find out. There is no other option. As you flip through the pages of *Homestead Tsunami*, you'll be faced with some hard facts, inspiration, and encouragement. A fire will most certainly be lit under you. You'll either walk away encouraged to start or continue your homesteading journey, or convicted of a consumer naysayer mindset that probably discouraged homesteading before reading this book. Whichever it is, you'll be changed and challenged. It's like sitting down with an old friend who isn't afraid to shoot straight with you. Or the older farmer who sits next to you on the front porch swing in the summertime, sharing all his wisdom. We all need those friends, and the best one you can have when it comes to this topic is Joel. I'm grateful he took the time to sit on the porch swing with us to instill his wisdom through this book.

We've reached an interesting time in society where you either get on the ark, or you keep treading water until you can't. It's either or. Which, by the way, has been our most recent "why". Watching economical and societal declines (not just financial) across the nation, and beyond, have solidified our yearning of a simple, sustainable life. We're creating a parallel narrow path alongside the broad path that society entices us to walk on.

The broad path tells us to just settle for mediocre food from chemically washed and injected produce and meat. It tells you that the "normal" is setting your kids in front of a screen all evening after they get home from being taught to sit in front of a teacher all day. The "normal" is not being inconvenienced. The "normal" is not knowing anything about sustainable living because someone will take care of you. But the narrow path? Well, that's quite different.

The narrow path says that a tomato grown in your own garden might take way more work but it tastes so much sweeter and is so much healthier. It shows us that even though having our kids at our elbows may feel like an inconvenience, that it's really a blessing for them, and us. It shows us that preserving food is hard work but it's good work. The narrow path says that being inconvenienced is actually a convenience, because we get to watch the next generation retain these skills and use them more easily. We raise children that know how to help us do the hard work. We create systems that work for us without the unwanted policies of man or government. We get to walk the narrow path, which means we get to live in true freedom.

We are a community that bears the Joseph mantle. In the book of Genesis, there is a man named Joseph. You could say he was a homesteader, but honestly, he was just a man of great discernment. A homesteader who saw the warning of what was to come, and decided to change the direction of his entire nation. He was anointed to make a difference by being the difference, bravely and without hesitation. While a seven year famine was ahead of his nation, instead of being loud and fearful, he prepared by creating storehouses full of grain and food. When the famine finally hit, not only did his nation depend on him because he was so wise and discerning, Genesis 41 says that all nations on earth came to Joseph because he had such an abundance from the seven years that he had meticulously cultivated a place of refuge. Eventually, even moving his family to the land of Goshen where there was an abundance and safety.

However, an interesting part of the Joseph story is when the people of the nations didn't prepare individually, their land and way of living was overtaken by pharaoh. While still being taken care of, they lost everything because of the ruling authorities, and their individual lack of preparation. Suddenly they were dependent on man and government, instead of taking heed and preparing their own households. I think it's time to prepare your household, don't you?

In a time that is as uncertain as these times we are living in now, there's no doubt about why there is a homesteading tsunami bearing down on this nation, and this entire earth. From America to Europe, from Asia, to South America— homesteading is now everywhere, in every land, in every way. Don't be simple minded enough to believe that homesteading is only about food. Quite the contrary. We are the movement of all things "home".

Homesteading. Homemaking. Homeschooling. Home-birthing. Home-growing, in every way. We are the generation and movement that has seen the corrupted societal village and know it's simply not the village we want ourselves or our children to grow up in. We need rich dirt on our hands from the garden that will feed our family for the year. And that dirt will teach us that hard work is good work. We need a milk pail full of warm, frothy milk from the milk cow that feeds ourselves and our entire farm. We need our children running through the grass, collecting flowers, bugs, and knowledge. We need a big dose of reality that we are generations removed from knowing real skills and now, we need them more than ever.

We need a disconnection from the rat-wheel and a plugin to the Creator who created us to do all of the things that bring this life back to the centerpiece of "home". The things that bring us back to the land of Goshen—a place of safety, refuge, independence, and abundance.

The beauty of it all is that it looks different for everyone. There is abundance on 100-acre farms, and there is abundance on quarter-acre homesteads. There is abundance on the prairie in the Midwest, and there is abundance in the coastal cities with rooftop gardens. Abundance doesn't just mean more than necessary. It means even more than you could've imagined coming from the space that you have, with the time that you're given.

As the wildfires of homesteading begin to sweep through your life, the heart begins to yearn for more and more. Maybe you're called to be on a rural farm taking care of your family and your extended family and friends. Or maybe you're called to be the Joseph in the middle of New York City—a beacon of hope and life. Whatever it is, wherever it is, I encourage you to pursue it wholly.

Homesteading is a back to the land movement, but it's even more than that. It is a movement that brings society and our culture back to reality. Where we appreciate our food and health more. Where we appreciate our spouse and children more. Where we appreciate the simplicity of how nature works in an ecological symbiotic relationship that keeps working just fine until humans mess it up. The earth, and everything thereof, groans for this reality. You, yourself, may even be nodding your head in agreement.

As you dive into the pages of this book, I know without a doubt that you'll be ignited, or reignited, to embrace the wave as it begins to crash down. In these pages you'll laugh, you'll cry, you'll say "yes" and "amen"! But most importantly, you'll close the book and say, "we have to get on the ark". Let this book be a testimony and encouragement to you as you set the ember of hope ablaze in your heart. Or, you simply reignite that fire that went out long ago because it all just seemed so pointless.

The wave is cresting. Are you ready?

Amy K. Fewell
Founder of Homesteaders of America
Author, Kingdom Builder, and Homesteader

Preface

By Joel Salatin

A merica is undergoing a profound homestead tsunami. Families are pouring out of cities seeking small acreages in the country. Podcasts speaking into this space have hundreds of thousands of followers.

Homestead conferences now fill my speaking schedule, and they're all over the country. You can find a homestead conference to attend virtually every month of the year. In the last couple of years, a steady stream of visitors, many in rented motor homes, have stopped at our farm to walk our fields and dream. Their most common reason: "we're escaping and heading for the country. We don't know where we'll land, but we're getting out."

A shaky economy, crime-ridden cities, fragile supply chains, empty supermarket shelves, increasingly invasive government regulations, dysfunctional mental health, kids addicted to social media—all these things make thinking people want to disentangle from the system. Stalwart American institutions, both public and private, are no longer trustworthy. Corruption, cronyism, and crisis screams from media headlines—or gets censored.

Wanting out when you feel chased and strangled is a strong incentive. An even stronger incentive is wanting something better. Wanting out and wanting in are two sides of the same coin. You can't flee without something to embrace. Obviously if you want out, the question is: where will you go? What are you going to do? You can't leave without going somewhere. You can't escape without a safe haven.

In 2020, 1 million backyard flocks of laying chickens germinated in America. Think of that. Assuming an average of six birds per flock, that's 6 million chickens. If they laid only 50 percent, that's 3 million eggs per day, or 250 thousand dozen. In a country of 100 million households, that's enough eggs to supply a dozen a week to a quarter of the nation's households. That doesn't seem like much, unless you're the only family with eggs. Or the family that can eat eggs for $2 a dozen when the stores charge $7.

In 2020, seed companies sold out. Canning lid inventories vanished. The number one Googled recipe in October 2020 was how to make sourdough bread. When the foundations of society crack, from political corruption to social media contamination, more and more people want to return to sanity and simplicity. Most people share a deep intuitive belief that if things go down, they don't want to be stuck in the city.

Throughout history, disintegrating societies de-urbanize as people head for the hills. As society collapses, you want to be near creeks, springs, trees, fields, wildlife (to eat when things get really tough) and away from attractive targets for bad guys, be they domestic or foreign. In especially dire circumstances, the countryside offers caves and coves in which to hide and disappear from chaos. Under political tyranny or social pressure, families who espouse non-mainstream beliefs and who want to protect their children from indoctrination seek solace in country solitude.

To be sure, not everyone can move to the country right now. Fortunately, we've never had as many gadgets and tools to grow things in the urban sector. I applaud every effort to engage with self-reliant living in the city. This book is not disrespectful to folks who, for whatever reason, opt to stay in urban settings. And I'm certainly not under any illusion that this book will suddenly collapse our cities. My whole objective is to offer reasons to make a change. Most won't. But some can and should.

Seeing the handwriting on the wall, many of us sense a historic inflection point in the world. When we see the World Economic Forum agenda, we tremble. This agenda is not a friendly agenda toward freedom and personal opportunity. Tracking your every move. No private property ownership. Government intervention in every aspect of life. People being fired from their jobs for not getting the COVID jab. Fake meat. Fake money. Defund the police. Teen depression and suicide. We can list plenty of things to be concerned about, even to be angry about.

This book is about taking all that frustration and anger and turning it from negative energy into inspiring positive energy. The objective is clear: when society becomes hopeless and helpless, some of us who build an ark will provide hope and help. We'll offer havens of protection and nourishment to lead our culture into stable families, fertile soil, nourishing food, working faith, and overall health.

Many families have already taken the plunge. They've invested in a homestead and are now in the throes of learning about chickens, weeds, and tomatoes. Many others are watching YouTube videos, listening to podcasts, and having long discussions deep into the night about whether they should pull up stakes and head for the hills. Others want to make a change but friends and family deride them, telling them they can't have a decent future living out with cows and pigs, beyond pizza delivery.

I'm writing this book to three groups of people.

1. The folks teetering on the precipice, trying to decide whether to jump.
2. The folks who jumped a year or two ago and are now discouraged because their visions of happy animals and flourishing vegetables turned into wayward critters and wilting cucumbers.
3. The folks who think all of this fleeing talk is nonsense. America is a land of plenty; why do you think things could go downhill? We've always been on top of the world. Live conveniently, and all will be well.

As I write, I imagine a representative of each of these three people sitting across the desk from me. Throughout the book, you'll see me address one of these people particularly from time to time. I'm trying to keep eye contact, make sure nobody misses the point, and bring the discussion along with relevance and clarity.

Homesteaders aren't normal. They don't whine; they get going. They don't wait for someone else; they take leadership. They don't compete; they share. They're interested in anything that makes life more self-reliant and independent from nefarious agendas. For the most part, they're enjoyable and productive neighbors.

Homesteaders don't buy fashion magazines; they buy journals that explain root cellars and herbal therapies. They don't go to movies; they build campfires by the pond and watch the moon rise on summer nights. They don't watch TV all evening; they can tomatoes and chase fireflies in the meadow. They don't spend all day playing video games; they gather eggs and fill their nostrils with the sweet aroma of fresh hay.

A couple of years ago, after being besieged by questions about raising livestock on a small scale, I wrote *POLYFACE MICRO: Success with Livestock on a Homestead Scale.* That was my how-to contribution to this wonderful movement. Since then, the movement trickle has turned into a tsunami. The first responders didn't need a lot of prodding. They'd always thought this way, and circumstances convened to push them over the edge. But the next wave mostly is coming with fewer roots in the thinking and foundation of homesteading.

These newcomers need a why. Many have zero rural experience, connections, or history. For the record, I'm thrilled with these urban transplants who have said "enough" and invested their nest egg in a country acreage. Many, many more need to follow. We need more people in rural America to make a critical mass that will keep the livestock, equipment, and feed suppliers in business. Industrial agriculture is killing authentic farming and land stewardship as much as food processors and bureaucrats.

This new generation of homesteaders is a shot in the arm for rural communities. Old ecological farmer geezers like me see these newcomers as the most exciting thing to happen in a long time. As a fulltime commercial farmer that doesn't use chemicals, who fertilizes with home-made compost, and believes pigs should express their pigness, I get no embrace from the conventional commercial farming community. On the other hand, most city folks generally love me but don't understand my world or want to immerse in it. That's okay. They make great patron saints to buy authentic food and keep me in business.

Oh, but these homesteaders. They get it. They're like sponges, soaking up everything they can learn. They've run away, yes, but more than that, they've run toward. They're embracing a new life. For them, as well as those who need to follow and those who don't get it yet, I've written this book to express the why of this modern homesteading tsunami. After letting me visit with you through these pages, I hope you'll either want to jump, rekindle your first love by being reminded of why you jumped, or know why your friends are packing. Now let's visit.

Homestead Tsunami

Good for Country, Critters, and Kids

Chapter 1

Get in the Game

When store shelves emptied during the spring of 2020, when the COVID Black Swan enveloped the world, did you panic or yawn? At our farm, we yawned while most people panicked. That's not said pridefully; it's simply a statement of fact.

Several hundred jars of canned produce filled our basement pantry shelves. Well-stocked freezers full of meat and poultry promised delicious, nutritious meals for months. Bulk flour, stores of maple syrup and honey, along with a neighbor's raw milk, assured us a future of thriving, not just surviving.

For decades, America more than any other country in the world offered a false promise that we could abandon historical participation in the foundations of life, somehow freeing us to do more important things. We would be free to spend more time keeping up with celebrity culture, going to movies, attending football games, and visiting vineyards. We could go on cruises, shop in Paris, play more golf, and hit the slopes routinely. The ultimate freedom—time to play more video games.

Liberated from life's chores and drudgeries, we could

really live. No more weeds to pull; no more cows to milk; no more hay to make; no more eggs to gather; no more tomatoes to plant, trellis, and water; no more meals to cook; no more canning to toil over in a hot kitchen; no more firewood to cut; no more larder to inventory and stock. We would all sail off into some sort of blissful Star Trek nirvana, eating laboratory concoctions, popping pills, and wearing spandex instead of Carhartts. We could unmoor from domestic drudgeries and escape into artificial intelligence, leaving our mundane lives free to pursue exotic activities.

Iconic radio commentator Paul Harvey differentiated between two types of freedom: the freedom to do what we want and the freedom to do what we ought. He likened the freedom to do whatever our whimsical desires wanted to a driverless car and a train without a track. They're free, but not in any sense that is functional. What does freedom to do what we ought look like?

It looks a lot like participating in life's most basic foundations. Over the last couple of generations, the hardscrabble American life, centered around home and hearth, has been replaced with convenience. We've contracted out the cornerstones of life to others, assuming new-found freedom would bring us to better places. Viewing domestic chores and basic life immersion as unnecessary slavery, as a culture we embraced TV dinners, cardboard tomatoes shipped across the country, Hot Pockets, and Lunchables.

In the name of convenience and liberation, we enslaved ourselves to a host of dependencies. We abandoned the one-room school with complete local control to a federal bureaucracy, quickly driving our nation downward in academic competency. If you look at what eighth graders in Laura Ingalls

Wilder's one-room school house needed to know, you'll see how far education has dropped. I have trouble with these tests, and I have a college education.

The historic in-home and locally-sourced entertainment and recreation model is now an outsourced video game conceived by nameless, faceless people a world away in both location and values. Instead of neighbors and friends getting together for a spelling bee, poetry sharing, or mock political debate, where everyone brings something, we sequester on our screens with someone else's imagination. Instead of a creativity potluck, we cater in entertainment, allowing us to slay virtual bad guys with violent weaponry in a virtual world. For many, we spend more time in a world that doesn't exist than the one that does. Is that healthy?

Instead of investing our money in local businesses and developing community-based entrepreneurship, we invest in Blackrock and massive global holding companies that use children to mine lithium. All to power our electric vehicles. Consider the care and attention our forebears put into the livery, keeping the horse fed, shod, and groomed. The early automobiles, built in backyard mechanic shops, could be maintained by anyone with a mechanical bent. Today we need a multi-million-dollar computer interface just to diagnose a problem, which is often some malfunctioning computer micro-chip sensor.

Rather than the neighborhood doctor who would often take a sack of potatoes as payment for a house call, we go to government-sanctioned hospitals that won't let patients take vitamin C supplements. Who wants to get tangled up in the medical system? How many of us lost parents during COVID, dying alone in hospitals and assisted living facilities, probably more from feeling abandoned than from any physical malady?

When the health care system gets its tentacles into your condition, your options become tormented by the clutches of a practically inescapable paradigm. The "informed consent" movement is a parallel tsunami to the homestead tsunami. They spring from a common root: loss of personal autonomy to make fundamentally intimate choices.

As a nation, we've abandoned Judeo-Christian moorings. Church attendance is at an all-time low, giving us more time to watch weekend sports, view pornography, smoke dope, seek exotic spiritual voyeurism, and a host of other miscreant activities. We took down the Ten Commandments from our school rooms and now wonder why we have school shootings. When I was in high school, the student parking lot was full of vehicles with guns—lots of students went hunting for a couple of hours after they got out of school. Nobody thought they were dangerous; we had prayer and posted the Ten Commandments in the classroom. Corporal punishment taught us early on that some behavior is okay and other is not. We called sin sin and evil evil.

Could it be that we traded the Ten Commandments for metal detectors and school security officers–and fear? Freeing us from those historical moorings brought us to a new non-freedom. Morality is now completely relative; no absolutes. Freed from biological constraints, our children suffer sexual dystopia, suicide, and self-hate. That sounds like a bad trade. The spiritual hole in the human heart must be filled with something. If it's not filled with truth, it'll be filled with untruth and nonsense. Some may call this the purpose hole. Call it what you will, but something about gathering eggs, milking a cow, and snapping green beans fills that hole with meaningfulness.

With all this going on, however, nothing indicates the dramatic societal shift as much as where we live and how we live, physically, on the landscape. Why plant a garden when we can go to Kroger? Why know a farmer when Costco is around the corner? Why run a freezer when take-out is an Uber Eats away? Cooking from scratch is now opening a can, pouring it in a micro-wavable dish and nuking it for a few minutes. Voila! Dinner is served. More households don't know what to do with a whole chicken. Butternut squash? What's that? Squash is supposed to be in a can.

All of this sounds progressive, cool, modern, and liberating—until some unexpected crisis like COVID hits the culture. When the supermarket shelves go bare and people realize that cities only have three days' worth of food in warehouses, this carefree, liberated, non-participating existence suddenly becomes vulnerable and fragile. The Chinese have a saying: "Plenty of food, many problems; no food, one problem."

As America, and indeed the world, convulsed with supply chain issues, people panicked. Disruptions were real and catastrophic. Industries exterminated millions of chickens and turkeys because the mega-processing facilities, employing thousands of people apiece, stumbled and faltered. Having 5,000 people working in cool, dark, damp conditions with lots of blood and guts is not a healthy environment in the best of times, let alone when a roving sickness spreads across the land. Raising animals only to throw them away in landfills and incinerators epitomizes a system enslaved to a rigid, nonadaptive paradigm.

Administrators, human resources departments, insurance companies and attorneys devoted untold time and attention to new distancing, quarantine, and medical protocols. Whole sectors in these plants suddenly shut down, making them unable to take the animals scheduled for processing. When Americans saw empty store shelves, the industry exterminated the animals because it couldn't get them processed. Notice I'm using the term exterminate, not euthanize, which is the word preferred by the industry and politicos.

For the uninitiated, euthanized is a euphemism for killing and throwing away. In the poultry industry, the most common way to kill a whole 15,000-bird factory house is foaming. They shut all the windows and doors, pumping in soap bubbles until all the birds suffocate. While this sounds horrible and tragic, imagine if you were tasked with killing 15,000 turkeys in a Tyson factory house. How would you do it? The word euthanized generally means mercy killing or putting an animal out of its misery. Since these were not sick animals, the proper word is exterminate. In some cases, farmers simply turned off the confinement factory house ventilation systems and let the animals slowly suffocate or succumb to heat exhaustion.

Farmers didn't get paid for those animals. The company owns the animals. Farmers bought the lie that if they signed a contract with the industry to grow the birds, the industry would handle feed, hatching, processing, and marketing–traditional areas where poultry farmers had participated in the process. Subtherapeutic antibiotics enabled thousands of animals to be crammed into these industrial spaces, freeing farmers from weather, predation, and meticulous management. Free to grow millions of birds in confinement, farmers didn't have to hatch, own, process or market them. But they got to watch as the exterminators came to foam their birds. See the irony here?

As these massive, centralized companies struggled to

keep their doors open, smaller counterparts continued to function almost without interruption. The question is this: if instead of 300 centralized industrial mega-processing facilities funneling America's food to supermarkets, what if we had 300,000 community-based facilities processing the animals and vegetables for American families? Would the catastrophic disruptions we saw in the food system have been as severe? The answer is obvious.

And what if, like in our house, nearly all Americans had a stash of food in their home larder? Until extremely recent times, the larder was where a household stored its off-season food. Most people today don't even know the meaning of the word. Before supermarkets, all that food was stockpiled in the domestic larder. At exactly the moment when people feared covid, the added stress of having nothing to eat further depressed their immune systems. The single biggest immune system depressant is stress. If for no other reason than to eliminate that stress factor in the future, people need to re-institute the domestic larder, ideally sourced from known provenance. When the specter of catastrophe looms, knowing you have several weeks or months of sustenance stored can allay a lot of fear, and keep your immune system functioning on all cylinders.

Of course, you can buy prepper rations, but that's like an IV for trauma. You don't want to live on prepper rations. You want to live on things you're used to eating as much as possible. While you can survive on an IV, no one wants to live long-term on an IV.

For decades government experts and industry gurus preached efficiency. These credentialed pontificators spoke condescendingly about farmers like us who participated in the breeding, growing, processing, marketing, and distribution of what we grew. We were labeled backward and unscientific. Goodness, I've been called a bioterrorist and Typhoid Mary for not vaccinating my livestock. Our farm has never purchased an ounce of chemical fertilizer, preferring carbon-based compost instead. For that, I've been excoriated by the conventional agri-industrial complex, accused of wanting to kill half the world through starvation. After all, we all know compost can't compete with chemicals. Just like real cows can't compete with lab meat. Today's orthodox narrative parades its agenda as a solution to everything when in fact it's a trajectory toward fragility, ongoing catastrophes, and ultimately shackling dependence.

Efficiency, convenience, and segregated labor dominated the food and farm systems. Naysayers laughed "how can you feed the world with pastured poultry?" Or "how can we maintain fertility with compost?" Farmers bought into the industrial notion, thinking they could once and for all escape participating in their own fertility program, their own breeding program, their own processing, marketing, and distribution programs. Freed to do one narrow thing, they borrowed money to build confinement animal houses, massive single-purpose equipment, and lines of credit for fertilizer and chemicals to grow crops.

What had been a highly integrated farmscape turned into a highly segregated farmscape. This freed the dairy farmer to just milk cows, the chicken farmer to just grow chickens, and the orchardist to just grow apples. Diversification onsite was archaic and obsolete; the modern farm resembled a factory more than a living entity. In the name of efficiency, farmers abandoned mundane things like manure management, stockmanship, and garden mulch. Replaced with chemicals, crutches, and convenience, devotion to efficiency seemed to be a winning model.

Until COVID. Until the supply chains fragmented. Until nobody wanted to come to work. Until Human Resources couldn't sleep at night, fearing a disgruntled employee would sue the company for inadequate health protocols. Centralized mega-companies spent billions of dollars in triage, trying to keep doors open, workers happy, and product flowing. That's when sourdough became the number one Googled recipe. Sourdough became the gateway to freedom. Isn't that the coolest thing? Few people had given any thought to it before. Suddenly, deprived of their consumer convenience and its false freedom, people by the thousands turned to a freedom based on participation. In doing so, they disentangled from the system and found new liberty to do, to try, to taste something created by newly-empowered hands in newly-appreciated kitchens.

Overnight, the business holy grail changed from efficiency to resiliency. Businesses realized that if you aren't first resilient, you have nothing to be efficient about. The first goal is survival, and in chaotic, dysfunctional waters, you don't want to be an aircraft carrier; you want to be a speedboat. Being nimble and maneuverable, making adjustments in real time, is the key to navigating treacherous shoals.

As supply chain issues reverberated through the food system, wild price fluctuations developed. Perhaps the biggest one was lumber. By the spring of 2022, lumber was up more than 400 percent, partly due to centralized mills' inability to keep employees coming to work and partly due to a wave of home remodeling projects. People stayed home, built that spare bedroom and fixed the deck, with bodies powered by sourdough.

Then, as if to add insult to injury, Vladimir Putin invaded Ukraine. Fertilizer prices shot up 400 percent. Ukraine couldn't ship wheat. Farmers lamenting exploding fertilizer prices filled news coverage. COVID bailouts brought on inflation, and suddenly Tyson announced a 32 percent increase in beef prices over 12 months.

A lady stood in front of our on-farm meat freezer and caught her breath.

"Are you okay? What may I do for you?" I asked her.

"Oh, I was just at Costco, and sirloin steak is $16.99 a pound. Yours is only $9.99 a pound," she responded, incredulity and shock etching her face.

At that moment, it all hit me. At our farm, we didn't need Russian fertilizer. We made our own compost. We processed our own chickens. We purchased our grain from local farmers, GMO-free (Genetically Modified Organism). In fact, back in the late 1990s, when GMOs first came in, we couldn't find a GMO-free alternative locally. We found a defunct 1950s on-farm feed mill in Stuarts Draft (a town about 15 miles away). The farmer eagerly brought it out of mothballs and put it to use. Today Sunrise Farms is the go-to GMO-free mill in the entire mid-Atlantic region.

What had our commitment to integrated farming brought? What had our diligent manure handling through composting brought? What had our dedication to on-farm processing brought? What had our love affair with often tedious direct marketing to neighbors and friends brought—all that inefficiency? In a word, resiliency. Suddenly the crisis enveloping the supposedly efficient industry showed its ends— fragility. On our farm, we didn't need to raise prices 32 percent.

More than a decade earlier, we'd invested in the only local federally inspected slaughterhouse when its owners aged out of the business. Communities all over America lost their abattoirs. "These small outfits are inefficient," all the Wall Street pundits and industry prophets said. These small plants were obsolete, according to the mainstream narrative. Like many others, it would have closed; we did the inefficient thing and preserved it. Again, this is not a boastful statement; it's an argument for resiliency rather than efficiency as a first goal. Certainly, efficiency is important, but only after resiliency is in place. Hundreds of these small 1960s community processing facilities went out of business as centralization promised economies of scale. Why invest in a small community facility? According to industry-speak, it didn't make sense. But then came COVID.

Suddenly our little abattoir, which had for years struggled to keep enough local volume to stay in business, was inundated with requests. People who had never considered selling a beef steer to neighbors had folks knocking on their doors asking for meat. Farmers began calling us by the dozens: "Do you have an available slot for me to bring in a steer?" For the first time in decades, we turned away far more business than we could handle. It was a tsunami.

During COVID, our little facility, never closed and didn't even slow down. With only 25 employees, spread out across numerous work stations, we didn't install plexiglass barriers and lose work days with quarantines. As far as I know, all small facilities like ours had the same experience. While the big boys stumbled and faltered, we kept right on chugging like The Little Engine That Could. Small community-based processing facilities like this offered a functional and resilient counterpart to the mega-industrial corporate facilities that exterminated and wasted millions of chickens, turkeys, and hogs.

In early 2022, with the war in Ukraine creating havoc in the grain and fertilizer sectors, our farm went through almost unscathed. Yes, we raised prices to accommodate inflation, but that wasn't anywhere near 32 percent. As I contemplated the sirloin steak story and all that was going on, I realized that our participation in what most farmers had abandoned actually freed us to survive and thrive during this dysfunctional time. Our farm had been accused of being elitist with our prices; now suddenly we were the most affordable place. Indeed, in the spring of 2020 we sold six months' worth of inventory in about six weeks. What a crazy time. As the aircraft carriers tried to adapt, to turn, our little speedboat kept right on going: the small abattoir, the small feed mill, and our carbon-based fertility program.

During this time, I received several phone calls from billionaires—that's not a typo—yes, billionaires asking, "how do I create an agrarian bunker for my family?" To be fair, the phrase agrarian bunker is my translation. They used phrases like "safe place" and "personal refuge." I knew what they meant. These were people who had private jets. They could go anywhere in the world, and they called me for an exit plan.

One even asked what "wheels fall off" meant to me. I discussed it with our farm team and we decided it meant you couldn't get fuel, electricity, or grain. While certainly other things could describe "wheels falling off," we opted for those three key elements. Then we realized that things are usually not as bad as you can imagine. One of the exercises for resilience is to imagine the worst-case scenario, then work your way back from there. As we analyzed our vulnerabilities, we realized that although we were not bulletproof, we would be the last guy standing. That position holds an optimistic promise because you always hope that by the time you're the last guy standing, someone will step up and figure out some solutions to the crisis.

By late 2021, a steady stream of RVs (Recreational Vehicles) came by our farm, piloted by families from New York, California, and other states looking for an agrarian bunker. COVID offered some positive developments for our country. One was accelerating the home office, reducing the in-person workplace and daily commute. Another was parents seeing what their children were being taught in school. And another was a renewed interest in food, rural location, and health. The one thing you don't want to be when the wheels fall off is sick.

Urban allure wanes quickly during a crisis. Intuitively, people realize that the city is not the place to be when the wheels fall off. Going to a place where we can see the stars at night, where we can collect rainwater, poop in an outhouse, grow our own vegetables and have a flock of egg-layers attracts folks who feel trapped by asphalt and technology. Urban living promotes places to go and parties to attend. Where you actually live is only a pit stop in life's race. But what if it's scary out there? What if nothing to sustain life exists out there? Stoplights and asphalt don't grow tomatoes. That realization drove billionaires to seek a place with soil, grass, water—things divinely placed and ordered beyond the grasp of humankind.

As a culture, we've spent decades abandoning the country, abandoning our kitchens, and abandoning connected provenance, only to find that the promised land of freedom is a self-inflicted enslavement to dubious agendas and fragile supply lines. We've chased a mirage in our thirst for freedom from chores, gardens, and kitchens. We've arrived at what we thought would grant ultimate liberty, only to find ourselves utterly dependent and shackled.

On our farm, we've built numerous ponds that grace the landscape with caressing riparian zones, providing habitat for copious numbers of frogs. Walking quietly and stealthily along a pond edge, you can often see the frogs before they make their protective jump into deeper water. I've never had a frog turn to me and say, "I'm not going to participate today. I'm just going to sit here and let the flies go by. I'm going to check out for the day." No, these frogs stay on point. Every single day they wake up and participate, just like everything else in nature. The leaves don't quit collecting sunbeams. The hawks don't quit looking for field mice. The buzzards don't quit looking for a carcass to clean up. The bees don't quit looking for flowers.

All of these things seem mundane. After all, look at all the flowers; who cares if one doesn't get pollinated? Look at all the sunshine; who cares if some rays hit the soil instead of green leaves? And to the buzzard, how about eating something else? Aren't you tired of carrion? Good grief.

But true freedom and functional thriving grow in mundane participation. It is in the milk cow, the fresh garden tomatoes, the hanging herb garden on the patio, the chickens devouring kitchen scraps and generously providing eggs for the family table. In the last couple of years I've realized that all the things the credentialed experts told us would bring freedom and happiness have actually enslaved us to a system. And it's not a benevolent system. Look at the agenda of the World Economic Forum and Bill Gates. They say half the world needs to be exterminated in order for humanity to survive. Food needs to come from centralized billion-dollar laboratories, not your back yard or pasture. What can people who truly believe this morally and ethically justify on their way to saving humanity?

Could anything be more dystopian? That is why more and more I hear people asking, "how do I disentangle? How can I build a parallel universe when the one being built for me seems completely out of whack?"

They feel trapped by the supermarket. Trapped by the public school. Trapped by their bank account. Trapped by their investment program. Trapped by their doctor. Trapped by family and friend expectations. Trapped by society's definition of success.

The recipe for freedom, for disentanglement, is not floating off into fantasy land where other people do the dirty work and think about how to get eggs on the table. The true recipe for freedom requires us to roll up our sleeves, engage, and participate in life's most basic and mundane tasks. Washing the dirty clothes. Cleaning the toilet. Butchering chickens. Churning butter. Cutting firewood. Picking apples. Canning applesauce. Yes, those homestead skills from a bygone era that created self-reliance and resiliency, that enabled people to live free from interference, that created liberty by embracing personal responsibility.

Doug Casey, quintessential libertarian and self-made millionaire blogged that the best hurricane shelter in a societal storm is a farm. Perhaps homesteads are the ultimate cultural hurricane safe place. Being free from the swirling chaos outside, snuggled with our family amidst gardens, orchards, cows, and chickens is the ultimate why of homesteading.

Revolutionary lifestyles create the path to freedom. Others can't do it for us. Although a lot can and should be done within the confines of the urban sector, nothing beats having a patch of land. Never have Americans had more opportunity or necessity to make the leap. For most of us, what holds us back is not financial or logistical ability; it's conviction. It's the why. Once we settle the why, everything falls into place. I've just articulated the real-time physical and cultural affirmation of a homestead mentality. Let's get on with it.

Chapter 2

I See a Homesteader

The term *homestead* only came into popular use in 1862 when the U.S. Congress passed the Homestead Act. The act had two primary provisions, both intended to settle lands on the western fringes of the nation.

The first provision allowed for anyone at least 21 years old and head of a family, either a U.S. citizen or intending to become one, to receive a title to 160 acres of public land if he lived on it for five years. While on the surface this may sound easy, it wasn't. The available territories lacked smooth-functioning services, from law enforcement to seed sales and repair shops.

The second provision allowed a settler to pay $1.25 per acre and not take up residence there. While this may sound like an incentive for moneyed people to buy numerous tracts and then sell them later, after development, such an investment would be considered highly risky. Investing on the ragged edges of society without living there made possessional clarity and authenticity impossible.

Back to the Revolutionary War, the fledgling nation wrestled with how to initially grant ownership to lands with no written record of ownership. The consensus at that time, disrespectfully prejudicial against Native Americans, was that the European concept of ownership and distinct property boundaries was the only way to ensure production. At that time, Americans did not yet realize that prior to European arrival, North America produced more food than it did then or even does today.

The Native Americans, just like all cultures, had some people who truly respected and understood ecology and others who exploited it. Some were more conservation minded and others took a more short-term view. For sure, fighting over resources and land control occupied much of their time. In general, though, their land management enhanced the indigenous ecology and it was highly productive. Some archaeologists and anthropologists believe Kansas and Nebraska, for example, had more people living there in 1500 than live there today.

Because the Native American-United States tension is still raw and is not germane to this book, I want to treat it and then move on. First, the whole land-conquering situation is complex and beyond my ability to take definitive positions. Native Americans for centuries fought over land, one tribe gaining ground and another losing, then later reversing it in ongoing wars.

Second, human history is a story of land control. Since history is usually written by the victors, it gives all of us skewed notions of the conflicts. Land conquest and loss permeates every acre of the planet. The actual catalyst for these disputes is often hidden from us.

Third, all of us, including those of us of European descent, can find conquest, servitude, and unfairness in our ancestry if we go back far enough. Human trespass and abuse frames civilizations as far back as people groups exist. Modern humans certainly haven't grown out of these issues.

Fourth, none of us can unring a bell. We can't correct every failure. We can't right every wrong. Life is rife with unfairness; welcome to the real world. Perhaps a sign of maturity is recognizing that forgiveness forges a better path than vengeance. Nothing can overcome hatred except love. True freedom comes when we release ourselves from vengeful blame. Every person can find unfair treatment in their lives. How we respond defines our life energy.

Fifth, I believe Americans violated far more treaties than the Natives. Had these treaties not been violated, we may have enjoyed a compatible peace. That's not what happened and our nation suffers from these deceptions yet today. But by the same token, humans deceive. We've been doing it since time immemorial, and we won't stop today. Nobody will ever live in a place free of deception.

Sixth, I can't fix it. What I can do is love those who are different than me. I can respect people with different ideas. I can shove a podium and microphone over to the person who argues with me. I can refuse to censor, hate, or question intent. If we all pursued this path, maybe we'd make progress. Now let's move on to the issue at hand.

As a civilization pushing its paradigms into a resource-rich area, the young United States saw settling these new territories as beneficial. Right, wrong, or indifferent, that was certainly the consensus of the nation. The primary debate was not that it should be done, but how. Farmers didn't want the land given away because that would depress prices if they wanted to sell. Northern states didn't want these new areas to be slave-holding regions, which is why the wrangling that tied up the homestead legislation during the 1840s and 1850s finally ended in 1862 with the secession of the slave-holding states.

Widely advertised in both America and Europe to attract settlers, the act soon fell victim to cronyism and add-ons. Initially, much of the land went to railroad tycoons. President Abraham Lincoln, who signed the act, also established the United States Department of Agriculture and gave some of the best lands to new government agriculture colleges. The settlers got the short end of the stick. Goodness, how things never change.

Between 1862 and 1900, somewhere between 400,000 and 600,000 settlers successfully fulfilled the terms of the act and became new settlers. Due to extremely primitive living conditions, nonexistent roads, and difficult transportation, many homesteaders failed. Most of them came from areas with more dependable and plentiful rainfall; the harsher midwestern conditions made successful crops few and far between. Local markets didn't exist yet. As a result, many turned into subsistence hangers-on, growing a wide variety of plants and animals to feed their own families, living an almost cashless existence.

Fortunately, the Internal Revenue Service had yet to be invented, and people actually could subsist on little cash. Through bartering and sheer determination, those who hung on until infrastructure and local economies could develop eventually thrived. Of course, many failed and their land either reverted to the public domain or sold to speculators.

In retrospect, the Homestead Act did not accomplish what its sincere-minded defenders intended—unless you're a conspiracy theorist and believe it was all a sham in the first place. Most historians now agree that homesteaders only settled one in nine acres of the available land. The other eight acres went to speculators, railroads, and big businesses positioned to control western resources, from mines to water.

But let's not let the final outcome jaundice our view of these visionary people who, in good faith, took advantage of the Homestead Act. Realize that initially these folks had to survive on the edges of society. They didn't have civilizational amenities. Even though 1860 technology was not as advanced as it is today, things were advancing rapidly. Steamboats plied the Mississippi. In 1837 Cyrus McCormick invented the reaper, which eliminated the scythe and is now recognized as the beginning of the Industrial Revolution.

Looking back, then, the Industrial Revolution was already under way by 1860. Many labor-saving machines were available. Railroads crisscrossed the nation, offering transportation options unimaginable only a couple of decades prior. Mechanized looms decreased fabric costs and home canning offered an alternative to smoking and dehydrating. While many of the modern devices we take for granted today— chainsaws, electric fences, four-wheel drive tractors—were not yet available, these folks left a lot behind in their quest for 160 acres.

They left behind friends, community, markets, jobs, and most of all, place familiarity. I'm writing this from my farmhouse desk in late December, looking out on freezing rain. We've just been a week without internet because ice coated the Wi-Fi transmitter that services our area. Nobody in the area has had Wi-Fi all week. You should have seen us trying to handle customers in the farm store. Fortunately, I can write this on Microsoft Word without using the internet.

My wife, Teresa and I don't have smartphones. Why? Primarily because we don't want all of our actions and thoughts entering a global algorithm, but also because we can't get cell service in our house. We're down in a valley in an American chestnut log cabin built in 1790; the walls are nearly 18 inches thick and don't let anything penetrate. Uber Eats does not serve our area. Papa John's won't deliver pizzas out here. The closest retail anything is eight miles away. If we want water, we can't hook up to a main; we develop it with wells, ponds, or cisterns.

Fortunately, we're on a main power line that services a prison about ten miles away; unlike many of our neighbors who spur off this line, we seldom lose power. That's a real problem in rural America. No whining here; just stating our reality. I've been told that fiber-optic cable is coming in the future; we see them installing it around the county. Maybe it'll come here someday, and maybe it won't. My point is that homesteaders today experience some of the same frontier-like infrastructure constraints confronting the settlers who headed west under the Homestead Act.

When I laughingly mention to people, "We haven't had internet for a week," they stare at me like I'm from Pluto. Let that sink in for a minute. No apps. No order ins. No social media. No "Google to find out." That's just as backward-sounding as not having access to blacksmiths, mercantiles, and taverns during the early days of the Homestead Act.

The type of people attracted to this varies greatly. Some are fantasizers living in la-la land. Somehow they think tomatoes grow by wishing them into existence, and cows want to stay home as much as the pet dog. While it's easy to poke fun, I still deeply respect these folks because they exhibit the courage to think differently. Even starry-eyed dreamers come

to homesteading because they value different things than
cultural orthodoxy.

Imagine the discussions among family and friends when
one of these adventuresome homesteaders announced, "We're
heading to Kansas under the Homestead Act." Most were
accused of being crazy, foolhardy, ignorant, and more. The
average person did not respond, "Oh wow, that's cool. Wish
I had the courage to do that." These homesteaders often
burned bridges to relationships, infrastructure, recreation, and
commerce.

For the most part, what drove them was yearning for a
better life. Deep down, they believed they could carve out a
better situation for themselves. They were willing to question
conventional wisdom, societal norms, urban sophistication, and
even economic consistency to embark on a new and different
path. Imagine the conversations between husband and wife,
wrestling with the various positives and negatives of what they
thought about but dared not voice publicly . . . yet.

What about this? What about that? What if? At that
time, these folks couldn't jump on a jet and return to familiar
havens if things didn't work out. This was for all the marbles.
It was for keeps. For many, they severed every security they
had and grasped their own security: perseverance, tenacity,
innovation, conviction.

With this in mind, I'm going to spend a moment paying
tribute to my own parents, who exhibited every one of these
homestead nuances and modeled them for me. My dad flew in
the Navy during WWII. After discharge, he attended Indiana
University, receiving a degree in business administration. My
mom attended Ball State University in Ohio. Having grown
up in a home with a deadbeat alcoholic dad (I never knew

my maternal grandfather), she developed an intense disdain
for alcohol in any form. Her single mom held things together
with domestic jobs, poorly paid. It was a tough life during the
Depression.

As a vivacious, smart, attractive college student, every
sorority wanted her to join, but Mom saw them as primarily
party outfits promoting the free flow of alcohol. She went to the
dean of students and asked permission to start an alcohol-free
women's sorority. Rather than congratulations, she received
reprimands. The dean wrote "troublemaker" in her student
resume because she wasn't content with what the college offered
and exhibited traits of rebelliousness. This happened in the
1940s.

Meanwhile, Dad, who wanted to farm as a vocation,
decided the new frontier was in the developing world and set
his sights on South America. He enrolled for a six-month stint
at Middlebury College in Vermont, known globally for its
language programs, and studied Spanish. Then he hitchhiked to
Mexico and lived with a family for six months before returning
stateside and sitting for the Spanish exam administered to
foreign civil servants. He passed and got on with Texas Oil
Company—later Texaco—as a bilingual accountant in the
fledgling oil fields on the coast of Venezuela.

He and Mom married and soon had enough money to buy
a 1,000-acre farm in the equatorial highlands of Venezuela.
They built a house and started raising chickens. With two boys
coming on, (my brother is firstborn, three years older) things
looked bright. Who needs groceries when you have bananas,
papayas, and pineapples in the front yard? But in 1959 a junta
brought the country to a standstill.

By this time, Dad had taken over the local chicken market

because our chickens were clean. The indigenous chickens, raised along village sewer gutters and under squat pots, all had nasal discharge (snot), indicating low-grade respiratory infections due to unsanitary conditions. Ours didn't. In the mid-1950s, in Venezuela, people didn't have refrigerators, and supermarkets hadn't come in yet. The food culture was primarily locally grown and sold. Farmers would take their wares to the village market, and vendors (middlemen) would buy and then hawk these goods through the city.

Customers would buy their food from these vendors. These vendors specialized in certain commodities: the banana man, the papaya man, the chicken man, the egg man, and so on. At that time, chicken prices fluctuated based on the health (nasal discharge) of the birds, which were sold live and tied upside down on a pole the vendor carried over his shoulder. Vendors wanted the best wares to get the best prices and develop loyalty among their customers. Before buying, ladies (virtually all customers were women) would run their fingers along the beaks of the chickens looking for nasal drip; the drier ones commanded the most attention and best prices.

Because Dad's birds had dry beaks, he quickly cornered the local vendor market for chickens, angering competitor farmers. They said our family engaged in voodoo and witchcraft to get such clean chickens. When lawlessness breaks out in a culture, it gives license and opportunity to settle scores that otherwise would simmer. When the junta developed, our family became prime targets for nearby farmers to settle the animosity, and we escaped out the back door as gun-toting thugs came in the front door. We lost everything. No amount of pleading would turn an ear at the highest levels of government. We left with a couple barrels of possessions and no money, landing in Philadelphia from a merchant marine on Easter Sunday, 1961. Dad was 39.

What would you do? Mom had a teacher's degree; Dad had accounting competency and Spanish language fluency. The temptation to just fit in, to go to the city and take it easy, had to have tugged mightily. Even to return to their Midwest roots (Mom was from Ohio and Dad from Indiana) where family and friends waited to embrace them and help them start over, beckoned mightily.

No, they decided to find another piece of land within a day's drive of the Venezuelan Embassy in Washington, D.C., in order to return immediately if things changed. I lost Dad in 1988 when I was only 31. The older I get, the more respect I have for his single-minded tenacity. When the future seemed bleakest, he persevered. He and we never went back, of course, and today we're thankful for this tragedy because it protected us from building something bigger that would have been expropriated in more modern times. I don't think Dad ever got over it. He left his heart and soul in Venezuela, loving the people, the culture, and the land.

His goal was always commercial farming as a vocation, not just a homestead. The difference is real, and I will do my best in this book to stay on topic and not veer over into the business of farming. Commercial farming is all about making a living from a farm. Homesteading is all about living securely on the land. While I do believe people can exhibit a homestead mentality without owning land, in its fullest application, property ownership is key to the foundational values of homesteading: food, fiber, energy, shelter, water, natural health remedies, and relationships. Perhaps we could put common sense in there as well.

These are not things to sell but things to do, be, and enjoy.
They are all things that precious few modern Americans know
anything about or even care about. But more people embrace
this every day. You don't have to sell anything. You try to
create sufficiency in as many areas of life as possible. You do
for yourself so you don't have to buy. Money you don't have
to spend is worth more than face value. If you save a dollar,
it's really worth about $1.40 because in order to spend a dollar,
you have to earn the dollar plus all the taxes associated with it.
If you don't have to spend a dollar, you don't have to pay taxes
on the earnings and you don't have to pay sales taxes when you
buy. These are huge savings. In practice, then, I try to live on
as little cash income as possible rather than trying to increase
my earnings. I'm loathe to use the word minimalist because
I don't consider this life minimalist except monetarily. True
richness doesn't come from money.

By June of 1961, Mom and Dad found this property
in Virginia's Shenandoah Valley, the most worn out, gullied
rockpile in the entire region. It was cheap. And 120 miles from
Washington, D.C. Only a visionary could have seen potential
in this property. Dad's father, my grandfather and namesake,
was a charter subscriber to *Rodale's Organic Gardening and
Farming Magazine* when it came out in 1942. In fact, for a
year the name was *Organic Farming and Gardening Magazine*.
Unable to gain traction with that title, J.I. Rodale switched the
gardening to first place and it took off. Lots more gardeners
than farmers. Still are.

Dad's eyes were set on full-time farming, and he consulted
with numerous private and public agriculture experts, posing
the question, "How do I make a living on this farm?" They
all advised borrowing money to build silos, plant corn, buy

chemical fertilizer, graze the woods, and buy equipment. Because of his dad's non-chemical approach, my dad eschewed all that advice and began searching for alternative options. He found Andre Voisin's *Grass Productivity*, which is still the Bible of managed grazing. Soon he designed and built a crude mobile electric fencing infrastructure so we could move cows from field to field.

We visited a farmer early on who built mobile shelters for his animals. I don't remember where it was, who it was, or what animals he had, but I remember Dad's enthusiasm on the drive home at discovering mobile infrastructure. Stationary didn't allow nimbleness; it was a simple business concept that resonated quickly with him. Within a year, he realized the farm couldn't pay a salary and mortgage, at least not in its current state and in his state of ignorance. We didn't have enough soil to hold up electric fence stakes. He mixed concrete in a wheelbarrow and shoveled it into used car tires, pushing a half-inch pipe in before the concrete hardened. He'd pile these up on the tractor platform and my brother and I, as little kids, could heave on the edge of these concrete tires and tip them off on the ground as Dad drove slowly. He'd then go back to stick electric fence stakes (3/8 inch rebar) in the half-inch pipes to set up paddock fences. Did I say this place was poor?

Within two years, Dad abandoned the commercial farming idea as undoable—for now—and took an accounting job with a firm. Mom still gets teary remembering how we didn't have enough money for Christmas presents for her three little children. Mom began teaching health and physical education at the local high school, and for the next ten years the off-farm paychecks gradually paid down the mortgage, and we reverted to a glorified homestead. I say glorified because it was

550 acres (160 open and 390 forested). We reforested 60 acres
of steep gullied hillsides, dropping the open acreage to just 100.
We planted acres and acres of trees.

The whole goal was to grow as much of our own food
as possible and drop our cash requirements. In 1973, Arab oil
producers cut off exports to the U.S. Known as the Arab oil
embargo, insecure and higher-priced gasoline added dimensions
to our homesteading. We abandoned the oil furnace and
installed a wood stove. That was a huge transition, not just
because of the money saving, but also because our own woodlot
provided our winter heat—this was solar energy. While the
wood stove wasn't as comfortable in the farthest reaches of the
house, it saved lots of money and allowed us to leverage wood
value, saving $3,000 a year in heating costs. Cutting that wood
was and still is some of our best income because we're saving
cash requirements. Today, we have an outdoor wood-fired water
stove that heats our house and Mom's house, saving around
$15,000 a winter. Add the taxes on that and it's a savings of
somewhere between $20,000 and $25,000.

The other thing Dad did at that time was buy a ten-speed
bicycle. Remember, Dad was now 52 years old and working
as the in-house financial officer for a metal fabrication shop
on the other side of town, about 17 miles from our house. He
began riding that bicycle to work and back, proclaiming that if
everyone would do that, the Arabs could keep their confounded
oil and we'll be just fine, thank you very much. Did someone
say something about thinking differently?

I could fill numerous pages with how Mom and Dad
thought and acted differently, but I hope these couple of stories
will help you understand that I stand on the shoulders of a
family that not only dared to think differently; they embraced

it. If one trait runs true through homesteaders, it is the desire to be different. We didn't plant corn; we built electric fences and installed ponds. We didn't buy chemical fertilize; we built compost piles. We didn't buy fuel oil; we cut wood. We didn't have a TV; we worked in the garden, canned vegetables, and churned milk into butter.

I grew up drinking raw milk like water—probably more than half a gallon a day. We had a couple of Guernsey milk cows and made yogurt, cottage cheese, and butter. When Teresa and I married, we remodeled the old farmhouse attic and lived there for seven years. When my grandmother (Mom's mom) couldn't handle her affairs, she bought a mobile home, put it in fifty feet from our house, and lived there for a decade, never entering assisted living.

After she passed, Mom and Dad bought a much nicer and bigger one and moved out of the big house, giving it to Teresa, me, and our two children. Who moves out of their big house before death and gives it to their kids? Living in that attic apartment (we called it our penthouse), canning hundreds of quarts of garden produce, driving a $50 car and wearing thrift store clothes, Teresa and I lived comfortably on $300 a month throughout the 1980s.

None of this makes sense to the average person and certainly not the normal urbanite. Teresa and I still like to see how little cash we can live on. We put money toward ponds, co-purchased a local abattoir to keep it operating, and re-started a local feed mill to guarantee local GMO-free feeds for our omnivores. That's where we put our extra money. Compared to most people scrambling for stock portfolios, retirement plans and fringe schemes like cryptocurrency, we'll take our larder full of home canned goods, earthworm-rich soils, and ten miles of gravity-fed water any day.

None of this seems valuable to the majority of people, and yet for those of us who embrace thinking differently, this type of living seems like the only sensible, reasonable thing to do. Whether we're in good times or bad, having some control of our basic needs reduces the need to worry and fear.

A homesteader, then, is someone who embraces a different place, always mentally and often physically, balancing running away from dysfunction with running toward self-sustenance. If you're consumed with worry and anger, homesteading will not solve your dilemma. If I asked you to make a list of all the things you're worried and angry about, you could probably make a long list. I could too. But if that's your motive for homesteading, you won't have staying power.

You need a healthy dose of optimism, of enthusiasm to embrace the different as a solution, not an agrarian bunker. The goal must be to take all that angst, all that negative energy that many of us feel right now, and channel it to a positive solution. As homesteaders, when the culture heads like a stampeding herd toward shallowness and dystopia, we must be beacons of hope and help.

We cannot offer refuges and solutions if we're thinking like others. If we're enamored of celebrity culture. If we have to see the latest movie. If our kids have to run in accepted social circles. If we're fettered to screens—of any kind. If we think an antiseptic asphalt world offers security and stability. If we wallow in victimhood and entitlement. Just like those Homestead Act settlers of the late 1800s, we need to embrace being different, go where others are afraid to go, do what others don't want to do, and lead by example.

Looking back, I view my formative years on this farm as

our homestead years. I think the farm might have generated enough income to pay the taxes, but with Mom and Dad both working in town and us kids too small to contribute a whole lot (I got my first chickens when I was ten), the judicious thing to do was put the farm business objective on hold. The experiments we conducted during those years, though, are invaluable today. They created the foundation for all our commercial innovations that today offer credible scale and community vitality.

We made mobile rabbit shelters, chicken shelters, veal calf shelters, lamb corrals, shademobiles, and fencing systems. We built compost piles, learned how to operate a chainsaw, and tried things that didn't work. Dad spent several years designing and building a loose hay dump wagon; after the first trial, we knew it wouldn't work. We tried artificial insemination on the cows; big mistake. By the time I came back to the farm full-time on Sept. 24, 1982, this homestead decade bequeathed tons of experience and experiments to leverage into a commercial enterprise.

Dad and Mom never made a living from the farm. My grandfather—the one who subscribed to *Organic Gardening and Farming Magazine*—never made a living from a farm, although he desperately wanted to. But all of this history provided a financial, emotional, and experiential foundation for Teresa and me to leverage as a dreamy-eyed young couple with a farm fantasy dancing in our heads.

Dear folks, we Americans—who developed the insecticide DDT, factory farming, McDonald's, and a food pyramid with crackers on the baseline—probably never embraced the Homestead Act, even when it was in vogue. But one at a time, one person, one family at a time, we're wandering back to the same character and conviction embraced by those hardscrabble, dreamy-eyed predecessors. That's who we are. That's what we do. Let's love it for its sheer compelling attractiveness in a fragile, topsy-turvy world. Let's be lighthouses of protection, steering our families to a more resilient way of living.

In the words of Robert Frost, "I took the road less traveled by and that has made all the difference."

Chapter 3

Filling Our Plates

How stable is America's industrial agriculture? Those grain harvesting combines you see in formation slicing across Iowa corn fields, filling bins and bushels with yellow starch are icons of American abundance. But are they vulnerable to anything? Self-propelled irrigation systems, known as center pivots, make round crop pies that adorn every aerial picture of far western cropland. But how secure are they?

COVID taught us to take seriously the musings of powerful global movers and shakers. When the World Economic Forum began hosting mock scenarios of a pandemic in 2019, most folks shrugged it all off as much ado about nothing. Until March 2020. Those mock press conferences and protocols became eerily real a scant few months later. I'm sure none of these global leaders had a clue that COVID might be just around the corner. The chances of them being that specifically prescient about a totally imaginary possibility is about one in a million. Mere peasants like most of us seldom really know what's going on. The insiders do. Sometimes they talk.

Recently President Joe Biden and mega-billionaire Bill Gates are musing about food shortages. Lots. One of the hardest things to do is getting people to see things that are not obvious yet. As a culture, we Americans assume we'll always have plenty. Even when we were consumed with World War II, nobody starved. That kind of stuff happens in other countries.

While, as a nation, we've escaped real food quantity shortages, the musings of powerful people about coming crises should give us pause. What could jeopardize our plentifully stocked and cheap food inventoried grocery shelves?

Here are a few things to think about:

1. Americans now import 20 percent of the food we eat.

We've never done that. Put another way, another country produced one in five bites of our food. Of course, I'm not opposed to imports and exports. But a big difference exists between food and agricultural commodities. Iowa, arguably the most productive farmland in the world, imports 90 percent of its food from out of state.

Cornflakes at a Kroger in Des Moines might be made from Iowa corn, but the main raw ingredient traveled out-of-state first to be made into cereal before being imported for consumption. The roughly ten percent figure for foods grown in-state, and never leaving the state before being sold retail to consumers, is fairly consistent across the food spectrum. Even in Hawaii.

The first time I spoke at a conference in Hawaii, I packed an empty satchel to bring home macadamia nuts. Teresa and I love macadamias and we figured they'd be cheaper there, where they grow. I'd check the bag of nuts and we'd enjoy them for a few months. When I arrived, I found no macadamia nuts. Production moved to Paraguay and Uruguay due to labor costs. In fact, I can buy macadamia nuts cheaper at my local bulk foods store than I could where they grow naturally.

Starting in the late 1990s, I began reading articles by forward-thinking experts decrying the low economic returns from farmland. Some of these college professors taught in agriculture college and still proclaimed that U.S. land was too valuable for the backwardness of growing things. They said we needed to grow data centers, Amazon warehouses, and hydroponic LED-lighted vertical food factories. We were far too sophisticated a culture to allow our precious land to be used to grow cows and beans.

I was speaking at a Xeriscaping conference in Arizona, and one of the other speakers headed up the Las Vegas water value project. Her team analyzed the value of a gallon of water growing onions in a farm field versus a gallon in a casino hot tub. Care to guess which gallon was deemed more valuable? The onion gallon was worth about 20 cents; the hot tub gallon outperformed it by $150. The obvious question is this: when push comes to shove, which one will win out in the public policy debate? That our nation employs people asking such a bizarre, myopic question and presenting their findings as a serious justification to forget about onions left me almost speechless.

The point is that our nation's dependence on foreign production is growing; it's never been this high before. We used to pride ourselves in feeding ourselves. No more. We pride ourselves in not feeding ourselves, instead putting buildings on all that worthless farmland and turning it into technology parks. That's the ultimate valuable and desirable use for land, after all. As this prejudice increases, we'll see more erosion in the production sector.

2. Soil depletion.

For a long time, mechanistic, industrial academics have disparaged soil as unnecessary to grow food. Soil is just a substance to hold up the plant, they said. We can feed it chemicals, called plant food, like an intravenous tube, they said. We can grow everything hydroponically; we don't even need the sun; we'll make sunshine in vertical urban factory farms, they said.

In the words of Dr. Phil: "How's that workin' out for ya'?" As nutrient measurement becomes cheaper and easier, we're learning that the "soil food web" brilliantly described by Dr. Elaine Ingham is important. The soil's 7 billion beings per handful are kissing cousins to our human microbiome; the two are meant to help each other. Only ten percent of these soil micro-organisms have names. Ninety percent are unnamed and their function is not yet known.

Leading voices that disparaged soil in the 1990s now realize it's important and the most valuable resource in any civilization. Yes, more valuable than the Dow Jones Industrial Average. More valuable than the Federal Reserve. More valuable than the Pentagon. And yet our chemical farming methods still deplete the soil far faster than it can be replaced.

Sometimes this movement toward appreciating soil's biology takes a step back, like the 2022 debacle in Sri Lanka, when starvation faced the country that overnight eliminated chemical fertilizer. It had chemical fertilizer proponents chirping with new confidence that organic farming is a recipe for mass starvation. If you're addicted to crutches, you can't throw them away all of a sudden, which is what Sri Lanka did. The production shortages there had nothing to do with long-term biological viability; the crisis simply affirmed once and for all that you can't shift from drug addiction (chemical fertilizer dependency) to clean living without some withdrawal pains.

If anything, the Sri Lankan example proved beyond a shadow of a doubt that chemicals and biologicals are extremely different production models. Perhaps the biggest lesson was how farmers who depend on out-of-country inputs to grow their crops are vulnerable to governmental stupidity. When a farmer leverages on-site carbon to feed soil biology, the dependency is on tried-and-true patterns and the access is under personal control.

New methods of mulching, crimping, gentle tillage, and pasture cropping offer hope for the future, but acceptance within the conventional agricultural sector is agonizingly slow. Drive across the nation's heartland in winter and you'll see mile after mile of unclothed soil eroding and baking. Chinese philosopher Lao Tzu said "If you do not change direction, you may end up where you are heading." Nearly every great civilization can trace its rise and fall trajectory to its land use. The Mesopotamians didn't think their fertile fields would give out. Nor the Greeks. Nor the Romans. Nor the Americans.

As Sir Albert Howard, godfather of scientific aerobic composting lamented in 1943, the temptation of every civilization is to turn into cash what nature spent a thousand years creating. Fortunately, with good stewardship, farmers can arrest and reverse the soil depletion trajectory, but we need to start sooner rather than later.

At least the Sri Lankans recognized a problem. Their solution did not follow nature's timetable and plunged them into catastrophe. Had they developed a strategic five-year plan to eliminate chemical dependency and rely on their own in-house carbon to feed the soil, the transition would have been beautiful to behold. American agriculture recognizes no such problem. Dead, chemicalized soil is just fine. Nutrient deficient food is just fine. Some of us want to escape this trajectory before it reaches bottom. Hence a homestead tsunami.

3. Water depletion.

I'm well aware of how fast things can change. A region devastated by drought, like the Colorado River Basin was in 2022, can suddenly be flooding within two months. Lake Mead could fill quickly if nature picked that area to dump on for a while. But anyone watching global hydration knows that marginal areas continue to grow. No real slowdown in the rate of desertification is happening globally.

Water allocation issues dominate politics in many areas, not just America's southwest and west. Morocco, the Sahel, and southern Europe are drying out. As a planet, we don't practice judicious water usage. Meanwhile, devegetation through overgrazing, tillage, and timber harvests change the hydrologic cycle. Instead of evapotranspiration forming healthy clouds, water vapors move and concentrate in far-off places.

Drought forces migrations. It even fosters war. Energy, water, and opportunity are the big three catalysts that move populations. Water is already in dispute on a large scale and will undoubtedly escalate. As more drought-prone farmland becomes idled in California, Colorado, and Nevada to make room for hot tubs, the easiest thing for rich Americans to do is import food from other countries. That's fine until the two countries have a spat. Or until supply chains strain under conflict between China and Taiwan or Russia and Ukraine.

The Ogalala Aquifer that underpins western agriculture has dropped about 100 feet in the last half century. While it

recharges less than an inch per year, irrigation sucks it down a
couple of feet. America cannot afford to deplete that aquifer,
but not only does depletion continue with impunity, most people
don't have a clue how to fix it. For one thing, we could quit
growing irrigated alfalfa to export to Japan to feed race horses.
We could return to perennial prairie and never denude the soil.
We could build organic matter for better water percolation.
We could do any number of things, but our agriculture system
whistles merrily into self-destruction.

4. Factory farm fragility.

Exalted as the pinnacle of American efficiency, the
Concentrated Animal Feeding Operation (CAFO) is one of
our nation's great contributions to the world. On the surface, it
seems successful, all those animals in such a tiny footprint. But
in reality, it skates on thin ice.

Probably the greatest success of the industry is keeping
catastrophes silent. Only the inner sanctum at Cargill, Tyson,
and Perdue, for example, are aware of massive losses. A
malfunctioning fan can easily kill thousands of chickens. High-
pathogen avian influenza seems to visit these state-of-the-art
facilities routinely these days. All the biosecurity measures
in the world can't overcome the basic fact that cramming that
many identical critters in such cramped quarters without any
natural habitat, sunshine, or decent air is an unhealthy idea.
You cannot create health in such an unhygienic environment,
no matter how many times workers walk through sheep dip and
shower in and shower out.

Animals are not far removed from people regarding needs
for clean water, clean resting places, and nutritious food. Just
imagine housing people like these animals are housed. Go

ahead, think about all the kids in your community housed like this, 24/7/365. To be sure, I'm not equating animals with humans; don't read into this any more than the point that hygiene is hygiene, whether it's a chicken or a child.

Once in a while a story escapes from the industry about a massive problem. Thousands of cows die in New Mexico feedlots during a heat wave. I had a neighbor who had 700 lambs in a repurposed factory turkey house. A cold snap froze the water, so he filled a tanker truck with water to deliver to the lambs. Unfortunately, he didn't properly flush the tanker, used primarily to haul chemical fertilizer. The poisoned water killed all 700 lambs. The media never heard about it.

What the industry presents as a fortress, a bastion of secure and stable livestock production, is actually one pathogen, one weather extreme, one infrastructure breakdown away from calamity. Are you going to put your faith in this tenuous system, or is something else more resilient?

5. Fertilizer costs.

The war in Ukraine brought the cost of fertilizer front and center. Anybody watching the news in 2022 realized conventional chemical farming had a big problem. With a war jeopardizing supply lines, farmers run the risk of being unable to procure the chemical fertilizers they need. Remember, industrial agriculture now depends completely on chemicals.

My dad likened it to drug addiction. As an economist, he viewed chemical use like a human drug addict. You have to use more and more, or get more lethal doses, in order to get the same reaction. Your body adapts to usage levels, so to get the same kick, you have to ramp things up somehow. This treadmill is inescapable until you completely quit the addiction.

On one of my trips to do seminars in Australia, I was standing in baggage claim after landing in Sydney. I have a big yellow cow eartag on all my luggage to make identification easy.

As mine circled around, I grabbed it, and two ladies standing beside me noticed the eartag. Most people don't have a clue what it is, but these ladies knew immediately what it was and asked me if I was a farmer. I said yes and then asked them why they'd come to Australia on that Los Angeles flight. They were genuinely excited to talk to a farmer and replied that they worked for a big herbicide company and were on their way to Perth for an international consortium to discuss superweeds.

They told me that, in Arkansas, farmers employ machete-wielders to physically go through fields and chop down superweeds that destroy sophisticated grain-harvesting million-dollar combines. These massive machines look rugged, but they are quite fragile if they encounter anything they're not designed to handle, like twelve-foot tall superweeds that look more like a tree sapling than a weed. I have a sneaking suspicion nobody left that conference believing compost and pasture cropping were an answer to their superweed problem. Herbicides created the superweeds through adaptation, and these folks confidently pursued more toxic, more expensive herbicides to deal with the problem their concoctions created. The average American doesn't hear about these things.

6. Energy.

Centralized production floats on oil. To be sure, energy is a front-and-center issue right now as one of the most significant arenas demanding innovation. Whatever the answers may be, at least for the foreseeable future concentrated industrialized agriculture, whether plants or animals, will continue to float on petroleum.

Here's why: centralized production knows no ecological boundaries. In other words, transportation limits that in earlier times constrained the amount and number of a singular species that could be grown in one place no longer determined concentrated scale. When transportation was slow and laborious, with draft power, you couldn't grow unlimited grain in one spot. It had to be eaten or processed nearby because hauling was too slow and expensive. And you could only grow it on fields near manure or other waste streams to maintain soil fertility. Prior to petroleum, all this transportation required animal power. Animals get tired. They can't go seventy miles an hour. They can't go on tracks pulling millions of pounds.

Expensive energy provided a backstop to human cleverness. You could only put as many animals or plants in a place as its ecology could support. Today, cheap energy overruns these historical constraints and allows for unprecedented segregation rather than integration. Farms used to be multi-species. Most grain went through livestock on the premises, and the manure blessed the fields with fertility. It was a beautiful, symbiotic closed loop.

Today, we've built a food system that eliminates the need to bring biomimicry to the farmgate. We can now buy our fertilizer from one place, apply it half a world away, to feed animals eating grain grown on the other side of the planet, to produce so much manure the surrounding area can't metabolize it. Everything that nature designed to be a beautiful and intricate relationship is now divorced and unrelated. Blessings are now liabilities.

Perhaps you think energy will continue to be cheaper. I

have no clue if we'll have another cheap fuel breakthrough. But if we don't, the energy costs of maintaining a profoundly disconnected and segregated production system will bear bitter fruit.

7. Commerce.

I'm old enough to remember prophets in the early 1980s forecasting a cashless society. These forecasts often ran alongside the development of computers and the internet, harbingers of a paperless world. Remember that? Computers and the internet now use far more paper than we ever thought imaginable.

Paper didn't leave, but commerce tracking is certainly rearing its ugly head. China leads the world in this technology, of course, but plenty of tyrants roam the halls of American government to implement draconian tracking options here. The U.S. Department of Agriculture (USDA—I call it the USduh) has long maintained a protocol titled Good Agricultural Practices (GAP). Sometimes these are referred to as Best Management Practices (BMP). They both mean the same thing. These are accepted procedures that form the conventional orthodoxy throughout agriculture. Canneries have these; abattoirs have them. Chicken farmers, pig farmers, and tomato growers have them.

Although they exist, they aren't universally mandated ... yet. But they are used informally to determine insurance eligibility, for example. In other words, if I want farm insurance and don't comply with GAP, the insurance company can say my exposure is too high, and I won't be able to protect my barn if it burns down. Vaccination programs, identification and tracking systems, many management protocols, and a host of things come under the GAP purview.

If you combine GAP with trackable digital currency, you can see how easy it would be to exclude unorthodox farming techniques (like pastured chickens and compost-grown tomatoes) from the market. And if someone purchased said unlicensed food items, they could lose their social investment score and be unable to borrow money or make financial transactions. I'm not trying to be a Henny Penny here, but as fast as some of these trends are moving, we have to at least consider them as part of food security.

The new horizon in this space involves mRNA inoculations in livestock. Genetic sequencing in vegetables in order to get them to carry various forms of vaccines is all being tested, with various timelines for rollout. None of this requires a label. For those of us who refused to take the COVID jab (remember, it's not a vaccine) the notion that a steak at Applebee's could send mRNA into our bloodstream is appalling and frightening. How about getting vaccinated by eating some broccoli? Folks, the American food and farm paradigm is completely off the rails. Commercial food is now entering a new phase of assault against biology and nature's templates.

The worst part of all this is that just like government thugs and society in general ostracized those of us who refused the mRNA COVID jab, these GAP protocols can determine what is legal in commerce. All for our own good, of course. All to protect us. When the powers that be decide that the only food that should be legal must be laced with protective mRNA, those of us who don't lace our vegetables and meat with these poisons will be in a pickle. And folks who want to buy unadulterated product will be in a pickle.

How do you opt out of orthodoxy when every transaction

you make enters a database monitored by gumshoes? None of us wants to think about these things, but serious people do think about them. We'd better.

8. War on meat.

Only if you've been living on another planet would you not be aware of the current war on meat. Ordinances have already been crafted to segregate meat-eating from non-meat-eating diners in restaurants. Burp and fart taxes on cows are not simply silly jokes anymore. Serious, powerful interests work on tightening the noose around livestock producers every day.

Bill Gates and friends are, of course, leading this charge, offering lab meat, fake meat, and anything but authenticity as the alternative to our livestock sins. Realize that the nutrition is nowhere near the same, even if it were as digestible. It's not. But the big thing to remember is that this represents the most undemocratic model for food sourcing imaginable.

With livestock, as long as the sun shines and the rain falls, you can grow herbivores without any other inputs. Omnivores are a little trickier, but as long as food scraps exist and nuts fall from trees, you can limp along with omnivores too. Contrast that with our food coming from multi-million dollar laboratories controlled by global industrialists; how secure is your sustenance?

That funnel can be shut off anytime to anyone for any purpose. A suburban backyard can feed a milk cow. Some food scraps can feed a laying hen and generate eggs, one of nature's most perfect foods. Realize, too, that lab protein requires massive feedstocks. Proponents make it sound like these labs create food out of thin air. "See our lab? It's clean and sterile and perfect for your Christmas dinner." They don't show you the train carloads of corn and soybeans going into the facility.

They don't show you the mono-cropped fields, eroding soils doused in chemicals to grow these feedstocks. Without livestock, the animal manures that have always been the bedrock of soil fertility won't exist. What will replace nature's perfect agronomic fuel? Oh, it'll be more chemicals, more superweeds, more petroleum requirements.

Where will the birds live? Where will the bees live? Where will the skunks, possums, and raccoons live? Documentaries using sophisticated video and audio techniques show pastures alongside conventional cropping fields. The absence of life in conventional chemicalized mono-cropped fields is obvious enough for a toddler to notice. The pasture buzzes with insects, birds, and bees. I can't imagine a more violent assault on our ecosystem than lab meat and all it entails.

But the biggest assault is on the universal access to food. If you run afoul of the powers that be in a world where food comes out of a mega-lab, you might be pinched off the provenance menu. At that point, you're going to wish you had a homestead.

Now we get to the point of this chapter: in a time of increasing fragility and centralized control, nothing provides food security like your own production. Everything you can grow around your home, in your home, in your yard—every morsel you grow yourself is one that escapes the control and agenda of a dubious corporate protocol.

This quick walk through the realities of centralized production should fire your heart to create a decentralized system. Nothing is more decentralized or more immune to the vulnerabilities I've articulated here than your own garden and family-scale livestock. The food you grow yourself doesn't have to be purchased or sold. It isn't in the commercial space. At least right now, it doesn't have to be registered with any government agency.

You don't need to comply with GAP or any other protocol. You're free to grow and eat the way you want. And when your food is walking around out your back door and growing in the front yard, you can see it. You know what went on it. You know what went into it. You know who handled it and how. You know the kind of protocols your food system rewards. You keep and leverage the value, not some nameless faceless mega-entity.

Securing the production of your food, to your table, to your family, is a compelling and legitimate why for developing your personal homestead. Of course, you can find local farmers to supply you. None of us is completely food independent. But even a farm like ours can't guarantee you provenance. If all you have to do is walk out your door and pick or butcher, you know where it is, how much it is, and its quality. That's peace of mind.

To drive the point home, imagine if hundreds of thousands of households embraced this idea. Our 35 million acres of lawn and 36 million acres housing and feeding recreational horses could be harnessed to break the stranglehold of an environmentally and nutritionally detrimental system. The cumulative effect of thousands of homesteaders on the market would break the pride and power of untrustworthy and fragile players. That's not just good for your family; it's good for the country. The greatest solutions are at the grassroots; your homestead is a piece of that solution puzzle. Don't worry about what you can't produce. Just jump in and pursue what you can produce, where you are, what you like to eat.

As thousands of us offer this alternative to our food system, it will bend away from where it is toward something better. Secure provenance is the place to start.

Chapter 4

Ready for the Table

E ven though it was half a century ago, I remember like yesterday a classmate's high school science experiment feeding one group of rats a name-brand boxed breakfast cereal and the other group all the raw ingredients listed on the label. The ones eating the cereal became weak and sickly; the ones fed the raw ingredients became sleek and vibrant.

Most of us don't have a clue what goes on between the farmgate and the cash register at the grocery store. Making Wheaties out of grain requires multitudinous manipulation. Remember the infamous "pink slime" in the award-winning food documentary Food, Inc. Stabilizers, emulsifiers, artificial sweeteners, and colorings change the nutrition and basic properties of food.

When our farm began making sausage, we naively thought we could concoct our spice blend, take it to the abattoir, and develop our own customized branded product. Guess what? You can't take your blend in a baggie to the abattoir, or at least not an inspected one. You can only sell meat that goes through an inspected facility. Any sausage blend arriving at that facility must arrive as a government-approved recipe in tamper-proof baggies.

In other words, we couldn't make a mix here at the farm and have the processor blend it into the meat. This required me to go searching for a licensed mix compatible with our philosophy. We researched dozens of blends and found four— yes, only four—that did not contain monosodium glutamate (MSG), a flavor enhancer. To this day, those are the four we use.

You know and I know that a huge difference exists between high fructose corn syrup and corn on the cob. I understand that food manipulation is a foundational element in culinary arts. But a big difference exists between arranging raw ingredients and treating them with cold or heat versus the unpronounceable chemical rearrangements possible only in a sophisticated industrial food laboratory.

This is why food writer icon Michael Pollan advises people who can only make one change to simply shop on the outside aisles of the grocery store rather than the inside ones. Fresh produce, fruit, meat, and dairy are always around the perimeter; the highly processed shelf-stable stuff occupies space along the inside aisles. But even the relatively unprocessed foods around the outside are subject to chlorine, other sterilization agents, and ripening gases; few people actually know what happens to food between farm and store.

The primary question to pose in this discussion is: Who do you trust? When you sit down with your kiddos to eat a meal, think about where you've put your faith that what you're ingesting is safe, nutritious, and good for the environment. We could even add good for the people. Large processing facilities are notorious for poor pay and working conditions.

When an outfit can't, won't, or doesn't hire folks from its own community, how socially responsible is it? Does the

labor that went into getting that food to your plate look like the
kind of work you'd like to do, or that you'd like your kids to
do? Is it a vocational option with broad appeal, or is it viewed
as undesirable? In Virginia, many years ago, the poultry
processing plants suffered financial losses due to workman's
compensation claims over carpel tunnel syndrome (repetitive
motion disorder). These thousands of workers did the same
action—usually with a knife—all day, every day. The industry
tycoons lobbied the Virginia General Assembly for relief and
received a nice exemption. By legislative fiat, carpel tunnel
disorder became a non-injury in the workplace, saving the
industry millions of dollars in claims.

I had a conversation with one of our Virginia state
senators during a particularly rough patch regarding
contamination in the large poultry plants, and he said the
biggest problem was getting workers to wash their hands after
going to the bathroom. Now you know why chickens have as
many as 40 chlorine baths before final bagging and labeling for
sale.

Interestingly, organic certification doesn't address this
issue at all. Animal welfare certifications don't address it at
all. Apparently, these certifications are more concerned about
pesticides than humans. Certification protocols cherry-pick
certain items to check and leave undone many more important
items.

Although meat and poultry are more my bailiwick, I
must point out that 95 percent of all food-borne illness and
contamination is not from meat, poultry, or dairy, but from
fresh produce. The primary reason is that it has no kill step
in the processing. Most other foods can either be heated or
supercooled to create what is called a kill step—a part of the

process that kills pathogens. But produce, and especially leafy greens, inherently preclude such a step. They can't be heated and can't be cooked. Their whole environment, from field to table, is cool, dark, and moist—ideal conditions for bacteria to proliferate.

As a result, chemical anti-microbial sanitizers like chlorine misters baptize the product during processing. During one of the last deadly illness outbreaks from leafy greens, I talked to a farmer who lived near the culprit processing facility. He said some glitch at the plant created a backlog of trucks hauling in the greens, and they couldn't keep everything chilled. As the leafy greens sweated in those trucks, all sorts of microbes took over and it overwhelmed the normal sterilization equipment.

If you're interested in this kind of thing, a bit of research will give you all you need to know to curl your hair. Again, I go to the question: Who will you trust? We all choose places to place our faith. As the food processing industry gained traction and the American consumer obliged with increased patronage, the scale, deception, corruption, and manipulation segregated food buyers from the processors.

Erecting fences and "No Trespassing" signs around these massive canneries, abattoirs, and food processors created an unprecedented opaqueness. Nobody knew what went on behind those guarded entry gates. In large part, we still don't. The paranoia that accompanies ignorance fueled the consumer advocacy movement, creating bureaucracies to provide what is euphemistically called government oversight. Of course, it is really the fox guarding the henhouse because the industrial food fraternity all go to the same colleges, attend the same social functions, and join the same clubs.

What started as protection for an ignorant consumer has now become a co-conspirator in the great authentic food delusion. Do you trust the CEO of Tyson when he says, "feeding you like family?" Do you trust the CEO of Driscolls when he lauds hydroponic strawberries? Do you trust the CEO of Nestle or Kellogg's or Danone, Unilever and Heinz? Really?

The food chain is just that—a chain with lots of links. When we talk about food, the discussion usually starts with the farmer, which is proper. But many things happen to that ear of corn, tomato, or chicken before it arrives in your shopping cart. As you look at that food, how sure are you that the process aligns with your beliefs? With your values? With your microbiome, for that matter?

A discussion about food security must include the processing portion because it is in this raw-to-retail state that some of the most egregious adulterations happen. If you can't trust that processing link in the chain, it doesn't matter that you trust the farmer who produced it and the transport company that delivered it to the grocery store.

The opportunity for shenanigans is immense. When our farm began supplying two nearby Chipotle restaurants with pork back around 2000, they supplied temperature strips to verify our cold chain. Unfamiliar with these nifty gadgets, I asked about their development history. Apparently, in the 1990s, long haul truckers began realizing that they could leave a meat packing facility in Nebraska, turn off their refrigeration unit, and save $300 in diesel fuel by the time they arrived in Baltimore or Philadelphia.

The temperature check upon leaving the Nebraska dock was right. The truckers would turn on their units in time to get everything cooled down enough to pass cold chain inspection at the receiving dock. But the meat spoiled quickly. Why? Internal investigations revealed the truckers' trick. To protect themselves, the industry invented inexpensive temperature strips that recorded temperature in timed increments. Tamper proof, these strips were put in some product at loading and pulled out at unloading. When plugged into a laptop USB port, the strip gave a temperature history to confirm adherence to the required cold chain.

Because you and I are honest, we have a hard time conceiving the kind of intrigue and shortcuts routinely enjoyed by large processing facilities working on slim margins with sympathetic government inspectors. One common trick is adding in-soak water to poultry in the chill tanks. Suffice it to say, Tyson sells a lot of water at the price of chicken.

With these stories in mind, I ask one more time: Who do you trust? As a homesteader not only producing your own provenance, but processing it, you have bulletproof assurance about the culinary manipulation of that food prior to your plate. Nothing else can give you the level of trust you can have in yourself, your own procedures, your own butcher table, your own kitchen.

Sometimes the processing step is completely separate from the production step. For example, Teresa and I, for years, have purchased apples from a non-chemical biological orchard about two hours away. In an annual pilgrimage, we go up there with boxes and bring home about a dozen bushels of apples in the late fall or early winter. For a couple of days, the kitchen becomes an applesauce commissary, and she cans 80 or more quarts of the most authentic, best-tasting applesauce you can imagine. It makes the grocery store brands—all of them, including organic—taste like junk.

Even if you don't have enough land to grow your own food, you can process it from known high-quality sources. You don't have to buy canned peaches. Every year, we buy several bushels from a wonderful Amish orchard an hour north of us. Again, the kitchen turns into a processing commissary, and the year's supply of peaches glistens in fresh jars placed on the larder shelf.

Why do we do this? First, because we know nothing weird or harmful adulterated the food. It's the real McCoy, as they say. Second, instead of being dependent on the industrial food supply chain, we already have the food on hand. Lots of people ask me how we eat, and my simple reply is we eat what's on hand. In other words, Teresa doesn't sit down and make exotic menus for the week.

Sunday we don't cook. We might heat up some leftovers, but mostly we nibble. An apple, some cheese, other fruit, nuts, glass of milk. Sunday is a rest day. I've always loved breakfast more than any meal of the day because I get up at daybreak, do chores, then come in to eat. Raw milk, eggs, sausage or bacon and fruit get us off to a good start. Long ago we quit a formal lunch because I was too unpredictable coming in for mid-day meals. To save the marriage and the kitchen, I consented to foregoing, nibbling, or leftovers of my own discovery—get my drift?—for lunch. Older now, Teresa and I have pretty much become a two-meal household. During the summer, a couple of freshly-sliced tomatoes and raw Gouda cheese is about as easy and satisfying as a midday repast can be.

Supper (or dinner, as some refer to it) is a wide assortment of soup, casserole, canned or fresh vegetables and sometimes bread or dessert (usually either, or). The garden yields corn, green beans, potatoes, tomatoes, cucumbers, asparagus, squash, beets, carrots, lettuce, peppers, cabbage, onions, rhubarb, sweet potatoes, strawberries and more adventuresome things from time to time. Beef, pork, chicken, turkey, and venison offer unlimited variety and abundance for the table. Surrounding the garden, grapes, apples, paw-paws, plums, blackberries, raspberries, pears, and mulberries offer seasonal freshness and frozen, canned, or dehydrated delight.

Anyone who thinks a homestead and eating from your own provenance is somehow limiting has obviously never enjoyed the variety and abundance offered in do-it-yourself food production. A cow and a few square yards of grain can add yet another layer of self-sufficient succulence to satiate your stomach. If you read that sentence out loud the Ss will almost make you slobber. I haven't even mentioned exotics like currants or gooseberries, peanuts and hardy pecans or English walnuts. Eating from the homestead is only limiting if we don't avail ourselves of its opportunities.

If we have a bumper crop of potatoes, we eat more potatoes. Do you know how many ways you can fix potatoes? If we have extra applesauce, we eat more applesauce. By the way, I can eat applesauce at any meal, any time, all the time. It's definitely one of my favorite foods. But the watery tasteless stuff out of the store? No thanks. You want your kids to eat better? Feed them your own processed food. Trust me on that.

The processing industry most folks find most repugnant is the animal sector. The filth and mistakes inherent in extremely large slaughterhouses are legendary. Animals skinned alive, chickens scalded to death—plenty of documentaries and books explain ongoing atrocities in great detail. And scale does make a difference. When you're processing 5,000 beef animals a day, cleanliness is much harder than an outfit doing 25 or 30.

Ditto poultry.

Blood and guts are much easier to handle hygienically in a small facility than a large one. Nothing beats your back yard. This is why home butchery classes are now one of the hottest staples at homesteading conferences. Because it's one of the most questionable industrial procedures, butchery is a great place to start for a viable homestead.

As a livestock farmer, I'm far more aware of dubious practices in butchery than I am in vegetable processing. I'm reminded of a joke about two guys on their way to a sustainable agriculture conference. One raised livestock and one grew produce. Getting hungry, they decided to stop at a diner for something to eat. Both were sensitive to food quality, but their hunger overpowered their trepidation about eating industrial food at a diner. Once won't kill you, they reasoned. When it came time to order, the livestock farmer ordered vegetarian, and the vegetable grower ordered only meat. Ignorance is bliss.

Few things can make you feel more secure than lying down with your beloved, knowing you've stocked your larder with delicious food processed under your own hand. It wasn't processed behind razor wire and security checkpoints. It wasn't processed by industrial entities with dubious objectives, like making money by adulteration.

You've overseen the whole process and can know for sure it wasn't tainted or compromised. If that isn't a good *why* for processing your own food in the homestead mentality, I don't know what is.

Chapter 5

Stashing and Stockpiling

❝ May I buy food insurance from you?" The young couple from Charlottesville, visiting our on-farm store for the first time, asked the question innocently enough, but it took me by surprise. The question caught me flat-footed and off guard; one of the few times I didn't have a response.

After I hesitated, they continued. "We're part of a 200-family group, and we're trying to guarantee ourselves food in case it's unavailable. We're looking for the same kind of security we have in other insurances."

Like you, I'd never heard the phrase before. Food insurance? We buy insurance for all kinds of contingencies: health, life, auto, homeowner. The whole idea of insurance is to protect you from some of life's biggest crises. I'd never heard of anyone applying that mentality to food, and definitely not in America, the land of plenty.

But here in front of me stood a 20-something couple, apparently part of a much larger group, who actually thought food was tenuous enough to require some sort of insurance guarantee. I assured them I'd think about it, and I've run the question by numerous friends smarter than I am to get their take.

The problem is that unlike all other insurances, which use money to pay for loss or calamity, when food doesn't show up on grocery store shelves, or if the food that's there is unhealthy and poisonous, the replacement food isn't readily available. If your house burns down, you can hire a construction crew with expertise, which then buys the electrical wiring, lumber, and plumbing to rebuild the house. But imagine buying fire insurance in a context where lumber was unavailable or construction crews were completely contracted out for two years. What good would your fire insurance do?

That question came to me more than a year ago, and I still don't have answers. The problem is that as a retailer (yes, I'm a farmer, and a retailer), I'm always eager to move inventory. Teresa's grandfather told me he had two rules for financial success for his farm:

1. "Everything is for sale."
2. "If someone comes along and wants to buy something at your price, sell it because they might be dead tomorrow."

Those two rules served him quite successfully. I well understood the problem this young couple and their friends faced. In the spring of 2020, when COVID turned everything upside down, three major changes occurred in our farm sales. First, we lost nearly all of our 50 restaurants. At the time, that represented about $750,000 in annual sales. For a small outfit like us, that was a devastating hit. If you had an eatery without a drive-up window, you were out of luck; we didn't service restaurants with drive-up windows. All of our restaurant patrons were dine-in outfits. Lockdowns were the single biggest economic boon to fast food in history. Thank you very much, government experts. As devastating as the COVID response was for most of us, it made a select few extremely rich. Fast food and pharmaceuticals, two partners in crime, benefitted the most.

Second, people flocked into the farm store who admitted they had never considered buying food anywhere except the supermarket. But these were strange times, and they went outside their comfort and convenience zones to get food. We had such a run on our inventory that we ran out of a few things. That did not bode well for our old loyal customers who had purchased from us for a decade and never been turned down. "What do you mean you're out of ground beef?" they asked. When told that all these newcomers had cleaned us out, more than one was irritated. "I've been with you for ten years; you mean you couldn't save some for me?" Of course, they didn't call and ask us to save them anything; they just assumed we'd never run out.

When our restaurants dried up, we were sure glad to have this crush of business. It helped make up the sales shortfall, and we were able to hold our entire team together. Even though we run a loyalty program for price breaks after X number of purchases, we never considered holding back a certain amount for those old-faithfuls. First, we never had to, and second, we could not know what they would buy. I suppose if we had a multi-million dollar sales tracking platform and personnel we could make a best guess, but we were flying by the seat of our pants and had no clue what tomorrow would bring forth.

As orders for volume purchases piled up, we realized we couldn't supply them all in a year, so we put a moratorium on big orders, things like half a beef and a whole hog. We love those big-volume sales, but we only had a certain number in our production pipeline and couldn't make a rapid adjustment. In fact, we had hundreds of pounds of ground beef in five-pound packages for restaurant accounts, but homeowners wanted one-pound packages. The truth is that holding back when someone is standing in front of you waving a credit card is almost impossible, both practically and emotionally.

The third drastic change was online orders. Fortunately, we had started nationwide shipping the previous July 4 (2019) and had worked out most of the glitches by the spring of 2020 when everything went into lockdown. Contactless retail took a quantum leap forward, and our online sales and shipping jumped by 300 percent for a couple of months. Once the supermarket shelves filled again, everything went back to normal except the restaurant sales.

All this change within our simple little farm business made me ponder this food insurance question. We had gone through massive sales upheavals in a couple of weeks. How could we possibly build something that would approximate regular insurance, with something like food? After all, we have a long pipeline. Let's say you want eggs. Chickens don't begin laying until they're five months old. So if you want more eggs, you have to set eggs to hatch for roughly one month, grow out the birds for five months, then go through the tiny egg start-up production phase for at least another month. That's seven months from the moment you decide you need more eggs. That's the best-case scenario, but in 2020 lots of things weren't operating on best-case scenario. Hatcheries ran out of chickens. Canning jar lids couldn't be found this side of heaven. Seed companies ran out of garden seeds.

What's the timeline for increasing beef inventory? If you want to increase your beef, you have to breed a new cow. Let's assume that's a heifer (unbred female). That heifer is two or three years old before she calves. Follow the timeline. Calf is born May 1. That yearling is bred in July the following year. After nine and a half months, she calves; let's assume it's a bull calf you'll convert to a steer sometime prior to weaning. The calf nurses all summer long and we wean the steer by Christmas. Then he grows another year and might be ready to convert to T-bones and ground beef by late the next year or early the following year. Have you lost track of the time yet? It's at least four years. That's a looooong lag time to expand our beef supply.

No short cuts exist. You can't sell what you don't produce. If we look closely at the food insurance concept, then, not only do we have the problem of a relatively inelastic supply, but we have the complication of perishability. Freezing does preserve for a long time, but quality will eventually deteriorate. In other words, we can't buy a freezer for our food insurance people, stock it with a pre-ordered guaranteed amount, and walk away. Somehow we have to turn over that inventory to freshen it up. Even if the turnover is twice a year, it's still a turnover. Compare that to ammunition, cash, or firewood.

While I've been describing meat and eggs, which do have a bit of wiggle room, imagine highly perishable things like milk, fresh produce, and fruit. The industry tries to overcome these inventory issues by picking green, in-transit gas ripening, and selecting less juicy cultivars. Even with all that protocol, processors and retailers discard millions of tons of fresh produce for various reasons. Blemish-free is the American mantra and it creates mountains of waste.

As I headed down this food insurance path, I realized that if I were truly going to guarantee meat and poultry to a certain group of people, I would need to eliminate as many vulnerabilities as possible. For example, for food insurance to kick in, the current industrial food chain would have to break. That means distribution and transportation would become dysfunctional. What could cause that?

Here is a list of things I can imagine; to be sure, I'm not predicting any of this. But if you're going to talk about food insurance like fire insurance, you have to start with assuming the building will burn down.

1. War.

Nothing stresses food inventories like war. Again, we Americans, having not fought a battle inside our borders since 1865, seem fairly cock-sure that the scenario themed in the movie *Red Dawn* won't happen. Bombs and prisoners happen over there, in other places, not here.

The Romans felt impregnable—until they weren't. Many times, wars are an internal conflict, like America's Civil War. Sometimes things are fairly localized, like the conflicts in cities over the last few years. Conflict that engulfs your area can make groceries hard to come by.

The average city in the U.S. stockpiles enough food for three days. In other words, if delivery trucks don't come for three days, people start into the Chinese saying of "no food, one problem." Nothing turns people aggressive like hunger. How long could your household stay put if armed conflict disrupted your food supply? Teresa and I recently read the story of the family who housed the Anne Frank family in Amsterdam during Nazi occupation. Not only did this family have to feed themselves, they also fed the Frank family. Many people starved in the city, but the ones who didn't, by and large, had connections to someone in the country who had a food stockpile. That is a lesson all of us should take to heart.

2. Civil unrest.

Just short of war or armed conflict, this happens from time to time. The Canadian truckers' strike in 2021 is a recent reminder that modern life depends on efficient and stable transportation. Look how fast governments intervene when railroads plan to strike. Automakers, not so much. But truckers and railroads, that's a big deal.

In other parts of the world, this is a huge problem. Local strongmen gain control of a bridge or section of road and demand money for trucks to pass. Smooth, stable distribution doesn't occur; instead, you'll get it when the truck driver negotiates passage through the barricade. Here in the U.S., truckers have stories about getting stuck in the wrong section of a city. Looters and hoodlums break into a truck stopped at a stoplight. After an experience like that, truckers tend to avoid those areas.

Good businesses and services leave areas overrun by lawlessness. Areas classified as food deserts are as much an indicator of civil societal breakdown as they are of any inherent food shortage. Again, plenty of food exists to feed the world; the only reason areas have shortages is because the food can't get to where it's needed, and that's largely caused by social/political upheaval.

3. Lockdowns.

The world has just come through unprecedented draconian measures to stop COVID. If you couldn't leave your house, how long would your larder last? Something as unthinkable as being unable to leave your house actually occurred all over the world.

Who could have imagined it?

In war rooms, people sit around all day dreaming up catastrophic scenarios and how to combat them. "What if" dominates the discussion. This is not paranoia, but healthy imagination leading to preparedness. Neglecting to think about negative what ifs is the surest way to succumb to predictable catastrophes. Better to plan and not need than to not plan and then need.

4. Grid breakdown.

I'm not a tinfoil hat kind of guy, but I also don't want to be a glib dismissive it'll-never-happen kind of guy, either. In recent years we've seen things like foreign hackers shutting down gas pipelines. I know that's technically not the electrical grid, but energy is energy and all part of the broader grid that powers everything. I was scheduled to drive to Tennessee during the weekend when the entire east coast was bereft of gasoline. I was glad we had a 500-gallon tank of gas here at the farm. The fear of being stranded on the road made for a harrowing experience.

And then, in November of 2022, someone shot a couple of electrical substations in North Carolina, plunging an entire county into darkness for a week. The perpetrator was not found. Why would someone do that? Petty mischief? Hatred toward a community? Hatred toward America? Vengeance toward society for some perceived unfairness?

Nefarious interests prowl our communities, unfortunately. *Mayberry RFD* is harder to find. I don't like to think about the vengeance that seethes beneath social veneers, but I know it's real. Every so often it rears its ugly head with some outbreak that makes us all realize stability and security occupy less time in human history than privation, conflict, and suffering. If the power went out, and your car was out of gas, and the gas station didn't have gas, what would you do?

Could you hunker down for several months with provisions you knew to be untainted? Asking this question and imagining these scenarios inevitably lead us to the *why* of creating a food inventory.

5. Sabotage.

Shortly after 9-11, the agriculture media carried numerous reports about sabotage vulnerability and how to make sure your farm was not susceptible. The most common attack scenario was poison in the water or feed. All written from the assumption of factory farms, these *what-ifs* imagined dead or tainted animals. They imagined vials of poison poured on produce in a spinach processing facility. They imagined bad guys entering a food warehouse at night and dripping poison into stored food.

When you think about how food recalls affect market behavior, just imagine what allegations of intentional sabotage would do. People would freak out. And rightly so. After the COVID response, we now know the capacity of fear to cripple and tyrannize a society. If some nefarious act made widespread fear of food a reality, the regulations, licenses, and government manipulation would escalate dramatically. Everything would slow down.

When these food/farm terrorism possibilities filled the agriculture media, I remember thinking how much safer our family's in-home domestic food inventories were. If someone wants to compromise a food inventory, they aren't going to do it in your house; they're going to find some warehouse, factory farm, some big outfit in order to affect a lot of people. Again, the smaller, decentralized food model comes out on top.

6. Disease.

Growing up, we never heard about *salmonella, E. coli, campylobacter,* high path avian influenza, and all the other maladies that have entered common usage. While diseases are certainly not a modern occurrence, the types we struggle with today are not the same as the ones we struggled with long ago.

Plant diseases are no exception. Contrary to popular thinking, we lose more crops to diseases today than a hundred years ago. Despite all the pesticides, fungicides and herbicides, crop losses as a percentage of total production are higher now than they were prior to chemical usage. To be sure, chemical fertilizer and genetic progress increased production per acre, but this production came at the cost of making weaker plants more susceptible to insect and fungal attack. While this is not a direct inventory problem, I include it here because some of the more fearful fringes in the ecological farming community say our narrowing genetic base is leading us toward an Irish potato famine future.

If that's the case, the food inventory is in jeopardy. Most plant geneticists now believe that had the Irish duplicated the genetic diversity still practiced by the Peruvian potato farmers, where potatoes originated, the famine would not have happened. But the Irish selected an extremely narrow genetic base that eventually succumbed to disease. Planting only one variety, and selecting seedstock from narrow parameters, created the disaster. In contrast, Peruvian farmers select diverse varieties with many different characteristics within varieties, offering broad variability and, therefore, ecological protection.

Heirloom garden seeds and traditional chickens have built-in genetic diversity to withstand the vulnerabilities that now plague the narrow-focused commercial industry. We find that kind of protection outside the back door, not inside the local supermarket.

7. Environmental disaster.

We've seen plenty of these in recent years, from Hurricane Katrina to Colorado floods to blizzards. Things happen. Environmental catastrophe is never fun to watch, but the human paralysis is often worse. Emergency personnel and volunteers who step up to the crisis offset the lack of preparedness that envelopes the average person.

Financial guru Dave Ramsey says everyone should have six months of living expenses in cash for emergencies. Life is punctuated with emergencies; if you haven't had one recently, count your blessings. Chances are the law of averages will catch up with you and you'll be in the midst of one sooner rather than later. The best way to mitigate emergencies is to plan for them.

Right now, the average American family cannot put their hands on $400 in cash. For a nation as wealthy as ours, that's ridiculous. But even more important than cash is food. You can't eat $20 bills. Or even Ben Franklins. I would suggest it's more important to have a couple months of food on hand than a couple months of cash. And it needs to be real food, not junk food. A freezer full of beef can sustain you; a freezer full of Hot Pockets, not so much.

Nature strikes with alarming regularity. If the grocery store floods, or the bridge washes out, what are you gonna do? Blame somebody? Shake your fist at God? Curl up in a fetal position until a firefighter carries you to a shelter? Our world is full of people who watch the game, fewer who play the game, and lots of people who don't even know a game is going on. Be one of the players.

My friend and fellow lunatic Del Bigtree said recently that we need to build an ark. Those of us who believe building an ark is wise shouldn't feel responsible for the folks who aren't interested. Leadership requires building and then getting on the ark. When the rain starts, dilly-dallying to argue with people who can't even see it's raining is foolish. We can't lead if we don't go to the ark. If folks want to join us, great. If they don't, sorry. Don't slow me down from getting to the ark; lead, follow, or get out of the way.

All of this brings us back to the original question about food insurance. I'm reminded of those first homesteaders who headed out to the Midwest under the Homestead Act. They purposely dedicated themselves to staying on their land. Homesteaders don't assume that when the chips are down, someone will come along and bail them out. As Teresa tells me, "You have to do whatever it takes to make it work." You can have your cry, your pity party, but only momentarily.

Then you dry your eyes, dust yourself off, and work where you are with what you've got. Why? Because you want to thrive when any or all of these tragic scenarios hit your neighborhood. Building an ark is not stupid. Noah endured 120 years of ridicule. Imagine building a massive boat on dry land without any way to move it. For 120 years. I'm glad the IRS didn't exist at that time to force him into a cash-based existence. Apparently, he enjoyed enough freedom to work on the boat full-time. Why? When everyone else drowned, he floated. The *why* of a food inventory: so you can float too.

Clear-headed people don't assume it'll never rain. We (I'm obviously including myself in the clear-headed group—ha!) assume it will rain. We assume one or more of these disrupters will touch our lives. Moving toward preparedness, in faith, dispels the fear and worry about *what if.*

If you're teetering on the edge, afraid to jump into homesteading, let this discussion be a kick in the seat of your pants. And if you did jump and you're now discouraged because fusarium wilt got the cucumbers, the neighbor's dog killed the chickens, and the cow aborted her calf, revisit this *why.* You're building an ark. You're building a nest, one straw, twig, and feather at a time. It all takes time, but as your nest grows and becomes more secure, you will enjoy deep satisfaction knowing you've invested in covering some bases.

When friends call you in a panic about something in the news, you can stand tall, stand into the tempest, knowing you have a fortress, a larder, an inventory of provenance to sustain your family. I'm concerned that too many of us put attention on building financial nest eggs, insuring our equipment, and even that our kids have similar experiences as their friends when instead we should be putting attention first on tomorrow's sustenance.

We should know where it came from. How it was prepared, processed, and preserved. We can trust it in bad times because we participated in putting it together in good times. Investing in that will return life-giving dividends. Why go to all that trouble? So you and your family can enjoy food insurance. The farther your food system moves away from conventional production, processing, and inventory, the more authentic, nutritious, and safe it will be.

Chapter 6

Sweating Together

My alarm rings at 4 a.m. I roll out of bed and go out to the chicken processing shed to fill the scalder. The shed, a simple pole structure about the size of a two-car garage and floored with a concrete slab, is a basic backyard chicken processing set-up. We're going to dispatch 200 birds this morning.

I flip the switch on the scalder to start the heating element, put on the plywood cover, and come back into the house. I reset the alarm for 4:50 and try to catch a few more winks of sleep. The alarm jolts me awake at 4:50. I nudge Teresa as I roll out of bed to re-dress. She gets up and within ten minutes we're out in the processing shed under lights.

Right before dark the evening before, I caught 100 meat chickens out of their mobile field shelters and loaded them in a trailer. Backed up to the processing shed, the birds have slept comfortably all night in their crates. Now I grab one at a time, loading it into the killing funnels (cones) and slitting the jugular. The red blood spurts freely, eyes close, and the chickens succumb peacefully and easily to the procedure.

While I dispatch the first group of birds, Teresa cleans

down the eviscerating tables with soap and water and begins filling the initial chilling sink. A used, tri-part stainless steel sink I'd acquired from a restaurant equipment sale works well as an initial chill tank before the birds go into their final chill vat. I check the scald water temperature one last time: 145 degrees F. Right on target.

I load the scalder with the first four birds, kill four more, and we're off and running. For roughly one and a half minutes the birds roll around in the scalder, soapy water penetrating the feather follicles and loosening them from their moorings. The scalder stops and I switch on the drum picker. It's a tub about the size of a washing machine with rubber fingers; a bottom plate spins, flipping the birds around and forcing their carcasses against the rubber fingers to pinch out the feathers. I spray in a dash of cold water and shut off the scalder. Defeathering takes only about 30 seconds.

I pull off the heads, cut off the feet, and put the birds on the evisceration table. Teresa grabs one and deftly guts it. I gut two and then return to the crates and cones at the beginning of our little homemade disassembly line to kill another group of four birds and keep the pipeline full. For two hours the two of us steadily work through a hundred birds. By 7 a.m. we're finished. I drain the scalder to add fresh water and restart for a second batch of a hundred.

Teresa goes into the house to get the kids up, dressed, and breakfasted. I jump on the tractor and head back out to the field to move chicken shelters and catch another hundred. I'm back home by 9 a.m. Our children, Daniel and Rachel, are up and ready for the day. I back up with the tractor and trailer of chickens. Rachel, the youngest, is just an infant. A playpen just outside the processing shed gives her a safe place but near enough to feel in the thick of it. Daniel is seven years old and big enough to help move carcasses and turn water on and off. Teresa and I duplicate our earlier routine, and by 11 a.m. we have 200 broilers cooling in the chill vats. We clean everything over the next two hours, which includes composting the offal and parking the tractor and trailer.

By noon we've cleaned up, grabbed some lunch, and then begin meeting customers at 1 p.m. Pre-ordered birds go out in household numbers as grateful, happy customers pick up their pastured chickens. Customers interact with the children. By 4 p.m., the chickens are gone, money is in our pockets, Teresa heads back in to prepare supper, and I head out to do chores. Daniel comes with me to help. We get back in by 6 p.m., eat a delicious supper from our own fields and garden and go to bed at dark.

If you're a homesteader, what I've just described sounds like heaven. If it sounds like hell, you probably aren't a homesteader. My most precious memories revolve around family and working together on the farm.

I remember in the 1970s when the agriculture media was abuzz about the farm of the future. Giant astrodomes would cover farms, creating perfectly controlled environments for everything. Farmers would simply sit at consoles and push buttons. In fact, the phrase futuristic agriculture prophets used was "push-button farming." Right after a flurry of these articles graced the pages of periodicals our family read, Dad and I were building a new boundary fence along one of our fields.

This particular piece of ground lay next to a creek. Some of our farm lies on limestone, some on shale, and in these areas near the creek, on what we call river jack. That's the kind of rock that forms the bed of mountain streams. They aren't boulders, but they are much bigger than gravel. While the ground surface next to the creek may look like good soil, within a foot the underlay is a packed jumble of river jack. What you see in the bed of the creek is what underlies the soil. Digging a fence post hole through river jack is no fun.

Dad and I were struggling on this particular hole where rocks nearly the size of basketballs needed to be pried loose with a digging bar in order to get down deep enough to set the fence post. The day was hot, and we were dripping sweat, pushing and prying, down on our hands and knees, extracting these rocks out of the hole. With all the mischievous humor he could muster, Dad looked at me from across the post hole and said, "This is some of that push-button farming."

We laughed heartily, both at our current circumstance and the arrogance and foolishness of the notion that, in the future, all of this would be accomplished from a push-button console. This was in the 1970s; this futuristic model was all going to happen long before the turn of the century. Well, we're pretty far past the turn of the century and we're as far away from push-button farming as we've ever been.

Interestingly, back in those days we never heard of things like high path avian influenza, African swine fever, mad cow disease and a host of other maladies now afflicting anti-natural agriculture systems developed by people who thought we'd have push-button farming. Today these same people say we'll eliminate disease with mRNA injections and replace farmers with Artificial Intelligence robots. Right before COVID, I was scheduled to debate someone at Oxford in the United Kingdom on the topic "The farm of the future will not have a farmer." I happily accepted the challenge to argue on the negative side, but alas, the trip never happened due to COVID.

Working together creates the best memories and builds stability in the family, which is the launch pad for all future relationships. Nothing creates close and lasting relationships like hardship. This is why soldiers have such enduring lifelong friendships. Something about coming through tough times makes enduring relationships. If you want good friends, don't go to bars where good times roll. Get in the trenches; dig a post hole; get dirty, sweaty, and struggle together.

Stable families create stable societies. It follows that if our society is becoming more unstable, our families must also be becoming more unstable. Said another way, if we want to arrest trends toward societal dysfunction, we need to put attention first on family stability.

Numerous trend lines indicate societal degeneration. Teen suicides. Drug use. Drug overdose deaths. Psychological problems among young people. Fentanyl. Opioids. Divorce. Homelessness. Domestic violence. Child abuse. That's enough for starters.

Along with the steady news stories about these negative trends, once in a while a study comes along with profound insight about how to change these trends. Several I've seen in recent years are:

1. The family that prays together stays together.

Regardless of religion or denominational affiliation, regular attendance at a house of worship and familial adherence to a moral and ethical code outside human creation reduces the risk of flying apart.

People who discard religious affiliations or reverence for the unseen realm develop a self-centered view of life. If I'm the center of the universe, I perceive things quite differently than if God, for example, is the center of the universe and I'm a player on His team. I won't belabor this point because this is not the area this book is about, but it's worth mentioning when discussing family stability.

Perhaps if we posted the Ten Commandments in America's classrooms, we'd have fewer shootings. Perhaps if we taught the Golden Rule as the bedrock of social function and civilized intercourse we'd have less strife. And perhaps if the average 10-year-old could still quote John 3:16 we'd have fewer teen suicides. The view of myself and my role is a direct reflection of how I view the greater cosmos.

2. The family that eats together stays together.

Numerous studies now show that eating together gives us a chance to decompress, share the day's experiences, and participate in communal emotional discourse. How much food today is packaged as single serving? Families graze independently, getting their single-serving food-like substance when the urge to eat beckons.

This hyper-individualistic notion that meals need not be shared facilitates isolation. On our farm, when we invested in a summer chef in order to eat the evening meal communally, it revolutionized the team dynamics. During the summer stewardship program (May 1-Sept. 30), we have roughly 25 people living here on the farm. Being able to end the work day with a family-style meal substantially improved the dynamics of our team.

It's a time of decompression, sharing, and planning. Eating together in an unhurried fashion gives us time to relax and recoup from the day's work pressures. As a farm business, we realized the value of this procedure based on the benefits

we knew from our family during the earlier years when it was
just us.

3. The family that works harmoniously together stays together.

Notice the word harmoniously. I don't want to get off on
a rabbit trail here, but a huge difference exists between working
together in tension versus in harmony. Families are rife with
conflict.

But I have a different take on family conflict. Rather
than giving up on the family, because everyone has its weirdos
and tension-filled moments, I submit that no place is a better
learning ground for conflict resolution. Because the family
creates our most intimate relationships and people are human,
after all, it will necessarily be the beginning of our most
frustrating conflict and memorable tension. But that doesn't
mean the institution of the family is a mistake; it means we
double down to make the family work.

We communicate. We call meetings, with agendas. We
hire facilitators. We enlist the help of counsel. And we're
patient. Patience to let things work out. Above all, we need to
believe the other person doesn't want to do me harm, but wants
to do me good. That forces us to listen, with respect. Not
cutting the other person off in mid-sentence. Yes, practice the
Golden Rule in real time.

Creating a harmonious work habitat requires mission-
agreed leadership—usually that's the parents. Having a clear
project list, a clear objectives list, and even clear spheres of
responsibility are foundations for an agreeable climate. A
homestead offers this in spades because it requires a lot of
different types of expertise at different times of the year.

An industrial factory farm, by contrast, tends to offer more singular projects. A Tyson chicken farm, for example, doesn't offer nearly the variety of tasks enjoyed on a multi-species homestead.

This variety enables each player to gravitate to an area of interest and expertise naturally. On a homestead, one person usually takes to the milk cow. That's a good thing, because cows tend to be one-caretaker partners; they don't like different sets of hands messing with their udders. Other family members tend toward gardening. Others tend toward culinary duties. Everyone can know a little about a lot of things, but usually individuals will take charge of duties in a natural, almost mystical, kind of good fit.

In our family, for example, Teresa drew the line at killing chickens. "I'll do anything but kill them," she told me when we first started. Rather than demean or disrespect her, I accepted it happily, probably because in true Golden Rule thinking, she was glad to do anything else, including gutting. We've always made a big deal around our house of never separating duties in a tier of sacredness to sacrilege. Everything is sacred. Today, we don't let our stewards classify work projects as "good jobs" or "bad jobs." If something needs to be done, it's sacred.

Another routine we started early on was enjoying ice cream treats after cleaning up from chicken butchering. We still practice this today; such a small thing, but that direct and timely reward for a hard job well done balances out work and play. When feathers and guts pile up, we know awaiting us at the end are ice cream snacks. What a change in attitude a little reward can bring.

Too many domestic living situations today don't require any work as a family. Not real work. When our daughter

Rachel went to college in North Carolina she was in a suite with three other girls in an off-campus apartment setting. None of the other three knew how to operate a vacuum cleaner. None of the other three knew how to take out trash. None of the others knew how to cook and marveled at Rachel's stash of home-canned meat and produce, asking, "What's that? What do you do with it?" I'm confident she was and probably still is the only student who arrived with more pounds of home-canned food than pounds of clothes.

A homesteader mentality permeates all of life. One way I know newcomers to the country are not homesteaders is when they buy fancy riding mowers for their lawn. Homesteaders relish work, enjoying substituting sweat equity for expensive convenience-oriented infrastructure. You know a homesteader moved in next door when you see chickens on the lawn, garden beds right up to the back door, and a hoop house instead of a garage. Homesteads have a utilitarian vibe rather than portraying a fashion statement.

One of my favorite memories of Rachel growing up was her "Lady Bug Club" with a couple of other girls. They created a newsletter to which the parents dutifully subscribed. At Thanksgiving, they each contributed an essay themed around gratitude. Rachel titled her essay "I'm Thankful for Work." She described how ugly the world would be without work. It's what keeps things functional. To say I was proud of her for that essay would be the understatement of the century.

A homestead requires a lot of work; an urban house doesn't. Few things irritate parents more than "I'm bored" or "I can't find anything to do." Really? A homestead affords countless opportunities to grow, build, and fix things. Some

families, of course, function better than others, but when all
are committed to respecting each others' gifts, no environment
creates more authentic opportunities to exercise relational
symbiosis than a homestead.

A close cousin of the homestead is a household.
As household businesses declined during the industrial
revolution, the practical need for families living under one
roof also declined. A family engaged in its business offered
opportunities for everyone to be needed. Dad had a craft; Mom
kept the books, interacted with customers, ordered supplies.
Children swept the floors, stocked shelves, and packed boxes
with merchandise.

When the Industrial Revolution took the man out of
the house and disbanded these family-centric businesses,
households collapsed because they had no reason to exist.
Husband and wife no longer worked together in their various
roles. Many argue this left women bored and feeling
unchallenged and even unnecessary. Children became liabilities
instead of assets. Youngsters without responsibilities and
expectations grew up being irresponsible and feeling entitled.
Maturity came later; instead of acting like men at 13, boys still
acted like fools at 20.

Things tend to break down when they're obsolete. In
many ways, modern American life renders the family obsolete.
With no practical glue holding things together and no reason
to feel needed, boredom, drugs, and screen time substituted for
historic relational and personal development. Our society is
now reaping this household obsolescence with a tragic loss of
purpose and self-worth.

I meet more and more parents today concerned about
screen time. Deep down inside, they know those hours a day

on a screen are unhealthy, but living in an apartment in the city makes policing screens harder. You can't even say "go out and play" in many areas due to crime and child safety. On a homestead, kids don't need any screens. After playing and working outside, they come back inside tired and head to bed. Nightlife isn't necessary on the homestead. How many parents worry themselves silly over their kids' prowling around at night? Put them under a cow at milking or weeding a patch of green beans, and I guarantee they'll be cured of being night owls.

On a homestead, lots of projects require a second pair of hands. Holding the other end of a board so one person can pound nails into the far end. Opening and closing a gate while sorting livestock. Harvesting garden produce always goes faster with additional pickers. Extremely small children can help tote firewood or pull branches into a pile.

Meaningful work makes meaningful life. Meaningful life makes us want to stay with meaningfulness. If that happens in a nurturing family habitat, chances are kids will always appreciate that formative experience. It can stabilize the family for years to come. I wish I could offer this as a recipe that always turned out right. It doesn't.

But life is about risk. I submit that the risk of family failure is higher when strenuous meaningful projects are not part of day-to-day existence. Something about weeding the green beans, gathering the eggs, and toting buckets of water, together, reduces the failure factor. We're still humans and subject to all sorts of dysfunction, but chances for successful stable families increase when we're co-laboring together rather than sprawling out in different parts of the house. If we've never learned the benefits of working together, we don't learn the benefits of doing anything else together.

We become in-grown, self-centered, and hyper-individualistic. When strong Mom or Dad comes along and helps kids who are struggling to perform, that's a memory of dependency; an affirmation that "we're stronger together than apart." You can't just preach that; you have to practice it. Extremely few jobs on a homestead are more efficiently performed singly; most are more fun and more efficient with two or more. The old adage "many hands make light work" is real.

One of the first things Dad and Mom added to our house when we moved in was a blackboard in the kitchen/dining room area. Installed in 1961 when I was only four, the blackboard is still there, and we still use it. It's now a permanent fixture of our family room, a place of instructional drawings, project lists, and artwork. When I was a kid, it held the day's work plan, which during the late summer always included chopping thistles.

My brother and I would spend hours chopping thistles, and since I was the extrovert storyteller, I'd maintain a running commentary on "The Salatin Boys Against the Thistle Nation." I'd explain how we were coming around the right flank and had the enemy on the run. I'm sure it bored him to tears, but I still have fond memories of chopping thistles together as kids. Why? Because we could see what we accomplished. Because we weren't alone; we were together. And because Dad praised us when he got home from work and saw what we'd done.

Meaningful work, accomplished together, is perhaps one of the biggest benefits of a homestead environment. We need never apologize for important tasks done well. Foundations for Farming founder Brian Oldreive has four criteria for defining meaningful, successful work:

1. On time
2. To standard
3. Without waste
4. With joy

A homestead provides ample opportunity to practice all these wonderful attributes.

While urban settings don't necessarily make lazy kids, homesteads can reduce the odds. Lazy kids often take a lazy view toward relationships, responsibilities, and commitments. Time spent honing work skills on the homestead tends to deepen all these stable-family requirements. The work facing us on a homestead should not deter us from embracing this kind of life; the work should be a catalyst to make us embrace it. Authentic family work is sorely missing from our culture today, and I believe it is a great loss in our techno-sophisticated, recreationally-oriented, pleasure-centric domestic environments.

Let's change the environment. Our homesteads can be vehicles to facilitate real progress in familial and societal stability. A lot of work? Yes. Worthwhile? Also yes.

Chapter 7

I Like Me

Am I valuable? If so, where is my value best actualized? What can I contribute to society, to family, to the world?

These are heady questions with profound practical implications. Affirming individual worth and feeling needed are primal requirements for emotional and physical functionality. Nearly every school shooting involves a person who feels worthless or unworthy. Whether it's bullying, poor grades, not fitting in or socio-political prejudice, these violent people, by and large, feel unable to contribute positively. In vengeance, retribution, or a final effort to be noticed, they commit horrendous violence against others.

As police investigate these incidents for motive, more often than not they discover a hodgepodge of perceived unfairness, being passed over, and hopelessness. These troubled young people finally come to a point where they don't think they'll ever be accepted, affirmed, or understood, and in a final act, go down in a blaze of ignominy. The patterns expressing this depression and vengefulness follow different routes between boys and girls. To my knowledge, not a single school shooter has been female.

But eating disorders are primarily female. Self-abuse is a kissing cousin to others-abuse; one is inward and one is outward. I'm no psychologist, but the findings I've read indicate that both stem from unhealthy personal metrics of self-worth. If a boy doesn't feel good enough, he tends to take it out on others. If a girl doesn't feel good enough, she takes it out on herself to achieve acceptance, attention, or audience.

If you talk to any school psychologist today, any counselor who works with young people, you will hear stories that tear your heart out. The maladies plaguing our young people these days are escalating dramatically. You can tell a lot about the health of a society by trend lines, and the trend lines for mental health among our young people are not good. Again, I'm not a scientist in this sphere but I do read widely because I'm deeply concerned about dysfunctional young people that become dysfunctional adults.

Further, I'm surrounded every day by a cadre of young people through our stewardship and apprenticeship programs, and personnel on our farm team. Unlike most farmers, I'm immersed daily in young people's lives and keeping this team, assembled from all over the nation, held together with mutual love and respect. It's not always pretty. Managing people is substantially harder than managing cows.

At any rate, the two things that constantly seem to bubble to the surface when dealing with unhappy young people are affirmation and expectations. My son Daniel says the quickest way to destroy a relationship is with unexpressed expectations. We can't read each others' minds, so if we want someone to know what we expect, we need to express it. How many break-ups have occurred because we assumed someone understood our desires, and it turns out we simply had different expectations?

This is not a psychology book, but my experience drives me to make an observation: we learn our self-worth through doing. We don't feel valuable because a focus group or self-help group made us feel valuable. Who we are is a direct result of what we do. Our thinking and actions go hand-in-hand; I don't know which comes first, but I know they are inextricably related. You can't do right without thinking right; and you can't think right without doing right and seeing good results.

This idea is certainly consistent with Jesus' teachings in the Sermon on the Mount, recorded in Matthew 6:19-21 (KJV):

"Lay not up for yourselves treasures upon earth, where moth and rust doth corrupt, and where thieves break through and steal: But lay up for yourselves treasures in heaven, where neither moth nor rust doth corrupt, and where thieves do not break through nor steal: For where your treasure is, there will your heart be also."

The point here is that if you start doing right, your heart will follow. Too often we pummel people with "get your heart right" instead of just encouraging them to do the right thing and their head will come around.

A close relative to the doing right is treatment of things. How do we treat things? Do we have anything for which we're responsible? Today's notion that children should have no responsibilities except to play video games and satisfy academia denies them experience in practical stewardship of physical things. If you're not responsible for a bank account due to having a business (yes, every 8-10 year old should have a business), or if you don't develop an appreciation for workshop or kitchen tools, you can't value things.

Children who aren't responsible for jobs or things don't value either and therefore never connect the value dots back to themselves. If I'm in charge of an activity or certain things like animals, plants, or tools, I learn to care. I care how the project looks when I'm done. I care about how the tomatoes look. I care if the tools work, like whether the ax is sharp or dull. Worth outside ourselves is the foundation for worth inside ourselves.

When children can't, won't, or don't have project and ownership responsibilities, they see themselves as worthless. Feeling needed and contributing to the good of family and society is the foundation for developing self-worth.

Don't think this is an excuse to be materialistic. I think having fewer things makes each item more precious. My argument here is not about how much we have, but how we value what we have. Even more important is what we do with what we have.

Here is my completely unscientific anecdotal old geezer observational hypothesis: accomplishment drives worthiness. The rest of this chapter builds on this idea. I'll start with my own life, sharing a deeply personal but transformative story.

Our family never had a TV. In the 1960s, as a kid, I didn't know about *Bonanza, I Love Lucy, Mayberry RFD, Gilligan's Island, Green Acres, The Ed Sullivan Show, Gunsmoke* or *Flipper.* I grew up on our farmstead (just a little bigger than a homestead—ha!), working and reading. I think I read every biography in our elementary school's library. I had a penchant for communication of all types: speaking, writing, reading. In sixth grade, I was the final contestant to go down in the school's spelling bee, making the first alternate in case one of the other three couldn't go to the regional bee. I was fairly smart, but not smart enough.

In the winter, after doing chores, I'd sit for hours at a little desk, writing stories with a Number 2 pencil in a spiral notebook. Just fiction. How I wish I'd preserved some of those, but they're long gone. I was extremely athletic until about 11 years old and then went into a pudgy stage. Suddenly I wasn't the first pick on the school sports playground. I wasn't the first one around the track; I wasn't the top chin-upper. But man, could I talk. In fact, my dad always loved sharing about his Parent Teacher Association (PTA) meeting with my third grade teacher. When he introduced himself to her, she said "Mr. Salatin, you have no secrets." Apparently I was a blabbermouth.

Remember that my mother was a high school health and physical education teacher, and she had been a stand-out athlete throughout high school and college. My older brother was extremely athletic, playing football and competing in gymnastics. Here I am, a chubby seventh grader surrounded by this athleticism, and I decided to try out for the middle school's baseball team. Looking back, I'm confident my decision to try out was to find affirmation and acceptance in my family first, and among my male school buddies second. To this day, I remember how I felt seeing the "who made it" sheet and my name missing. It stung.

About that time, the school had a forensics meet, and I competed in public speaking—won it. Then in English class the teacher decided to do a formal debate unit and needed volunteers to perform. I loved it and did well. The Daughters of the American Revolution (DAR) had a "Why America is Great" essay contest and I entered. Won it. At home, my chicken enterprise put money in my pocket. Adults who encountered me encouraged me in my business and thought a teen entrepreneur was pretty cool.

Are you seeing a pattern here? The school system abandoned the middle school concept for a couple of years at that time, which sent me on to high school in eighth grade. The sports program included three basketball options: varsity, junior varsity, and eighth grade. The eighth grade team played the other middle schools in the area. I tried out. I can remember like yesterday looking at the postings of the select members and seeing my name glaringly absent.

In that moment, I resolved to abandon formal sports. Backyard pickup football games, sure. But nothing formal. And I resolved to focus all my energy where I clearly had talent and success—communication and farm entrepreneurship. I remind young people today to be thankful for their failures because they help you determine your life's path. If you never failed at anything, how would you determine your best avenue of contribution?

The other point in this story is that while I was trying out for the basketball team, I had already joined the high school debate team. Although the Virginia High School League rules prohibited me from competing in formal tournaments since I was technically not a high schooler, I nonetheless threw myself into the team with the older students. I participated with reckless abandon. In our intra-squad scrimmages, even the well-heeled seniors didn't want to debate me. They loved putting the research off on "the kid," and I dutifully fed them material, but it was secondhand. When we practiced, I tore them up because I'd done the firsthand research. At least, that's the way I remember it.

The takeaway is that I was poking around many different things to find my niche. By this time, my laying hen business on the farm began to take off. I needed more outlets for my eggs, and we joined the local Curb Market, a Depression-era local market that by the early 1970s barely survived. What had been nearly 100 local vendors in the heyday of the 1930s had dwindled to two elderly matrons. And me.

They mentored me in marketing, customer relations, product display, and pricing. There again I thrived and added our homestead product line to the venue: butter, yogurt, cottage cheese, beef, pork, rabbit, chicken, eggs, and garden produce. I put in a vegetable garden completely separate from the family's and sold produce there throughout high school. An indoor, year-round market, it opened at 6 a.m. and ran until about 11 a.m. every Saturday. From age 14 to 18, roughly four years, I was up at 4 a.m. every Saturday morning, 52 weeks a year, to get my stuff together and be ready to meet customers at 6 a.m. The mentorship of those two matriarchs and that early interaction dealing with customers, was as valuable as anything I learned in school.

The homestead afforded me an opportunity to develop a childhood business, interacting with customers, honing entrepreneurial marketing savvy, and competing in an adult world. My final two years of high school, I worked the night reception desk at the local newspaper, writing obituaries, police reports, and answering the phone. I wouldn't trade that for a Ph.D. Video games did not yet exist and teenagers received welcome opportunities to participate in real adult vocations. Today, we call that exploitation. I say kids growing up on screens without meaningful things to do and stuff to own are being denied the most foundational opportunities to know who they are and why they're worth something. I'm not in favor of a Dickensian child labor exploitation, but I do believe the pendulum has swung too far in the opposite direction.

The point of this brief autobiography is that self-worth develops from being successful at meaningful things. It doesn't develop out of some ephemeral feel-good cloud. Of course parents need to encourage their kids, but valuable praise must be tied to something practical. You can't just say "you're a good boy" or "you're a good girl" and expect self-worth to sprout legs. Accolades are impotent unless they follow action. And the more meaningful the action, the more valuable the affirmation.

Our daughter, Rachel, began baking pound cakes and zucchini bread when she was eight years old. At the time, we participated in the recently started Staunton Farmers' Market, providing her an initial outlet for her craft. Sophisticated matrons would come by the farmstand and make a fuss over Rachel. "Oh, you're the one that baked that exquisite pound cake I served my garden club ladies on Thursday. We all just raved about it, and I told them I'd gotten it from you." Do you know what that did to Rachel's self-worth, her self-esteem?

Compare that to an affirmation completely detached from deeds. Our self-worth is tied up with activity. Our culture has done a great injustice to young people by criminalizing meaningful and practical work-place interaction with adults. We've called it child abuse and exploitation and relegated our young people to TV and video games. Being the top points getter on Angry Birds or whatever the game of the day, is a far cry from being a great butcher, baker, or candlestick maker.

Our nation segregates its young people from practical adult work interaction and then wonders why twenty-year-olds have such a hard time launching. They can't launch because they never built a launch pad for themselves. Goodness, they don't know what kind of rocket they are. Or if they are a rocket. Lack of direction plagues young people because they never went anywhere meaningful. Vacations, Disneyworld, ski trips, and cruises are all fantasy; they aren't real; in fact, they're designed to not be real. They're promoted and billed as getaways. But if your whole life is fantasy, what's the point of another one? A getaway as a reward for work well done might have merit, but otherwise it's just silliness on top of silliness.

Down in the trenches where life happens is where you find out who you are, what you like, where your talents lie. The Clifton Strengths-Finder personality/vocation template is based on the idea that rather than focus on overcoming our weaknesses, we perform better by leveraging our strengths. How do we know what our strengths are? By actively participating in an eclectic blend of meaningful pursuits starting the younger the better.

I had my chickens, garden, school extracurriculars surrounding communication competition, Curb Market, and working part-time at the newspaper. No TV, no video games; my life was consumed with meaningful activity which helped me determine my interests and talents. By the time I was 17 or 18 and had bloomed into a non-pudgy muscular physique, I knew I wanted to farm fulltime for my life's vocation.

I'll never forget my last visit to the school guidance counselor as a rising senior. She asked me what I really wanted to do with my life, and I unequivocally responded, "be a farmer." She went into hysterics. "What? Waste all that talent? Waste those brains? You're an honors student; how could you squander that on a farm?" It was a beating I'll never forget and one of the reasons I remain adamant about children having their businesses, staying off screens, and embracing working alongside adults in meaningful tasks.

You don't find out who you are, what makes you tick, your interests and talents by playing video games and hanging out. Listen, we've got a planet to steward and save. We've got earthworms to encourage. Pollinators to proliferate. Ecological protocols to innovate. Who has time to hang out? Part of the reason young people want to hang out all the time is because they've never embraced a cause big enough and compelling enough to consume their interest.

A homestead developed and maintained to secure food, build soil, hydrate the landscape, protect the commons, and defund the agri-industrial-chemical-fake food complex is a great mission. It's a great cause. To maintain physical health, mental acuity, spiritual stability—all the whys and greatest objectives of life—this is what a homestead offers. Entertainment and recreation don't give us purpose and direction in life. Part of our responsibility as parents and adults is to present our youngsters with a mission and vision big enough and sacred enough to leverage their creativity and life's energy.

I'd much rather be partners with someone who knows how to grow a succulent tomato than knows how to win on Game of Thrones. Intuitively, young people know the difference too. But most never have the option or encouragement to experience the more meaningful choice. They muddle through pre-teens and then teenage years addicted to screens and whatever social media demands. I can't believe the number of young people I encounter whose life's objective is to become a millionaire on YouTube or a sensation on TikTok. Have people done it? Yes, but it's a tiny fraction of the population. And what's the ultimate value? Does it build soil? Does it make us healthier? Does it heal the broken-hearted?

A lot more people eat eggs. A lot more people need clean water. A lot more people want great-tasting carrots. If the wheels fall off, do you want to be dependent on TikTok, or do you want to know how to build a compost pile and grow cucumbers?

When our children were extremely small—Daniel was 8 and Rachel was 3—Teresa and I received a conservation award and the reception required us to be gone for a couple of days. At the time, we had a herd of about a hundred cattle that we moved every day. I set up the electric fences for the paddocks and we left the children in my mom's care. For two days, Daniel moved that herd of a hundred by himself. Do you know what it does to the self-worth of a little boy to watch that herd respond to his best imitation of Dad's call, "Come on, Coweeeeees!" by gathering them up calmly to walk into the next paddock?

When we returned from our quick trip, all was in order. Daniel had taken care of everything . . . at eight years old. To be sure, I'm not suggesting that a child without this level of competency is a failure or even that being able to do this is a litmus test of success; I'm simply pointing out that a homestead affords unprecedented opportunities for children to accomplish things, meaningful things, for which Mom and Dad can praise them. While I'm not opposed to athletics, I'll dare to question if Little League compares at soul-depth affirmation with growing a tomato or moving a herd of cows.

Any household has chores or activities that need to be done, that are meaningful and therefore valuable in praise. But a homestead, in addition to the household, adds complexity and dimension that inside work doesn't offer.

Too often, our kids don't know who they are because they've never been anything. Oh, they've been places and occupied space, but what have they been? As adults, we greet each other with "what do you do?" Our worth is inextricably linked to what we do. If the response is, "I'm a bank robber" we have quite a different perception of the person than if the response is, "I'm the editor of an e-magazine." The desire to do something meaningful burns deep within the human heart. Nobody should work for money. We should work for causes. Homesteading as a cause is valuable enough, sacred enough, and even eclectic enough to capture our attention for a lifetime.

When Virginia gubernatorial candidate Terry McCauliffe ran for governor the first time, he convened a roundtable of agriculture leaders to solidify his interest in farming. Strangely, I was invited to that meeting. As the day approached, he promised attendees a special guest. About twenty people arrived at the appointed hour and McCauliffe, bigger than life, addressed us, primarily extolling the virtue of USDA grant programs and how his administration would ensure Virginians took advantage of every federal freebie possible. Partway through the summit, the big reveal happened, the surprise: Agriculture Secretary Tom Vilsack walked in. Yes, the head guy at the USDA.

Of course, to the McCauliffe campaign, this was a slam dunk. Getting a federal cabinet-level person to meet with your prospective constituents is a big deal. Vilsack began a rambling address about how important agriculture was and how we needed to preserve farmers, and then he let the final shoe drop by saying farm kids made the best soldiers, and that's why we need to keep farmers farming. I don't think he saw my jaw drop, but I've never forgotten it. When I blogged about it later, the McCauliffe campaign threatened me with violating their confidence (no one signed confidentiality statements when we entered the meeting), and said that if government agents hassled us, his office certainly wouldn't help, yadda, yadda, yadda. In other words, "don't ask anything from us, baby." He won the election.

Don't let the sideshow detract from the point of the story. People in the highest offices of the land believe farm kids make the best soldiers. Vilsack explained they are more creative, confident, and competent than city kids. Of course, when I blogged about this, urban folks were more upset at being told their kids were second-rate than the farmers who were cheapened into cannon fodder. For sure, I'm not suggesting homesteads are good in order to keep a steady flow of warriors flowing into the Pentagon; I'm simply pointing out that this notion that homestead skills create a sense of self-worth, a sense of knowing who you are, that "God don't make no junk" is not brand new in my thinking. It's axiomatic in the heady discussions of powerful interests and insiders.

As our child segregation (that's what I call "Prohibition to Work") laws become more strict, homesteads offer a return to yesteryear when children's lives integrated with the adult world. When we admonish kids to "grow up" we need to provide them a germination tray, an opportunity to grow up. You can't grow up playing video games and watching TV in the basement. Nobody in our day better articulates this than Professor of Animal Science at Colorado State University, Dr. Temple Grandin, one of the most remarkable people I've had the pleasure of befriending. Leveraging her autism throughout life makes her the world's leading animal handling facility designer. She can see what animals see. She can feel what animals feel. Her favorite refrain about young people on the autistic spectrum is to get them out of the basement and into the adult world, fixing cars, grocery shopping, mowing lawns. She literally preaches how necessary meaningful activity is to provide a sense of self-worth, a healthy mental state and a sense of purpose in life.

"The cows are out!" is one of the most poignant statements on a homestead. They could be on the public road or eating the sweet corn in the garden. Cows being out is never a good thing. Rounding them up requires finesse in herding. It requires a strategy about where you want them to go. It requires planning the direction and path you need them to go. It demands immediacy—you can't lollygag along when the cows are out. Learning that all our plans and expectations can, and indeed need to be, displaced from time to time is part of human maturity. When it's all over, and everybody laughs or cries or both, a deep sense of accomplishment washes across the team. "We did it. We got them in." It's primal: man, beast, elements.

What a difference between that and successfully ordering pizza delivery. Visceral activity and work with a purpose combine to yield awareness of who we are and what we should do. We can take all the personality tests in the world, go to all the counseling sessions we can find, and listen to motivational speeches, but authentic accomplishment is still the secret sauce to discovering who we are and what we're made of. A homestead offers an eclectic array of opportunities for that kind of discovery.

Proximity to births and deaths. Playing in the creek. Squatting along the pond's edge to watch tadpoles, salamanders, and dragonflies. Discovering the litter of kittens hidden discreetly in a crack of the haystack. The magic of sprouting seeds. The agony of fusarium wilt. The succulence of a first ripe mulberry. The payoff of abundant strawberries in a well-tended weed-free bed. Bringing in the harvest. Eating canned garden produce in the winter. "Remember that huge cushaw squash? This jar is from that squash, and we're going to make a pie with it." These are not just nostalgic memories; they are recipes for life's treasures and competencies.

If I asked you to list your child's skill set, what would the list look like? "Excellent with the TV remote." Be honest, now. Think. If you asked your child to peel potatoes, would he or she need any more direction than that? Gathering eggs. Splitting wood. Packing a wheel bearing. Setting up an electric fence. Even children who will never be farmers look back with fondness on these life experiences because they shape character and steer young people into productive adulthood. You will never feel better about yourself than your own competencies.

Mothering baby chicks is a great lesson in mothering children. Pruning apple trees is a good lesson for administering discipline to your kids later in life. The sense that "I've got this" flows discreetly but surely from these real-life homestead experiences. If we want a society full of confident, competent, caring adults, a homestead offers the best launch platform.

Chapter 8

Bright-eyed, Bushy-tailed Kids

In the modern American home, how many chores do children do? If we made a list, we'd be hard-pressed to find many. Let's see:

1. Make your bed.
2. Feed the cat/dog.
3. Take the trash out to the curb.
4. Wash the dishes.
5. Vacuum the house.
6. Wash and fold clothes.
7. Fix meals.

In chores, I do not include putting away your clothes, or personal hygiene like brushing your teeth or bathing; those you do because you're human. Chores are repetitive requirements that make the overall household function, not personal bodily care. I think I've been fairly generous with this list because I daresay in many households, several items on this list aren't expectations for anyone under 18 years old.

How do we teach responsibility, dependability, resourcefulness and a can-do spirit unless we're constantly put in situations requiring all of these characteristics? In the previous chapter, we drilled down on self-concept. Now we're drilling down on basic character—what others know you to be.

I like the saying, "Character is who you are when no one is looking." Our society is facing a crisis of accountability. Think about the arguments in the home over screen content. Think about the heartbreak when parents discover what their kids are watching. A constant tug-of-war exists between screen access and screen accountability. Censoring platforms constantly try to keep up with clever circumventions on the part of nefarious content developers. Circumvention is the name of the game, and it's tearing families apart.

For the record, I've never played a video game and don't have a smartphone. So I confess it's easy for me to be judgmental about kids, social media, screen time and content: that whole issue. Fortunately, our kids are grown. But my heart breaks for the temptations and pressures in families with kids, seduced by screens in all their permutations. For the record, I don't think smartphones are sinful or evil; I don't have one yet because it's not necessary and that's a way I can disentangle from the system. That's something I can defund. No cell phone works in our house; why have this expensive a gadget if we can only use it when we're gone from home?

I'm grateful I don't instinctively reach for a smartphone whenever I have an idle minute. I can think, dream, take in the surroundings. Flipping out the smartphone every time a lull occurs in life is a terrible distraction. We miss a lot when we're constantly engaged with our screens.

Well, what are we going to do about it? Fuss, fume, and frustrate? How about offering a more vibrant distraction to the smartphone? How about offering something more attractive than screen time? Enter the homestead therapy and child development program. Chores are the backbone of character development because they can start incredibly early in life, and they define the fundamentals of functional living. A household that struggles with the chore list I mentioned above is profoundly dysfunctional. You can tell a lot about a family's character, discipline, and vision by the way they keep house.

A homestead extends that domestic platform to a larger venue. Of course, parents must model chore diligence first; sloppy parents usually bring up sloppy kids. Conscientious housekeeping engenders pride and discipline, a protocol that extends into character development. Decision-making is a muscle that we can either develop or let atrophy. Chores place us into decisional roles from attitude to performance.

In the stewardship and apprentice program on our farm, we teach what the military calls situational awareness. As you walk around and do things, what do you see? This is why we don't allow screens out during work hours. You have to lift up your eyes, tune in with all your senses, and engage in your surroundings to cultivate this sense.

For example, when you approach a herd of cows, what are they doing? Are they placid or agitated? Moving or still? Grazing or lounging? Facing east or west? Lying down or standing up? Spread out or bunched up? Swishing tails or standing loose? Looking toward you or looking away? Is their rumen side filled up or showing a bit of unfilled cavity? Is anyone in heat, indicated by riding? If one is riding, is it a steer riding a heifer or a heifer riding a cow? This is all part of mastery in herdsmanship, but it starts with chores.

When you see the herd every day, you naturally begin noticing things. All the kinds of things I just mentioned. You can't shortcut this learning. Becoming a master takes repetition, and that requires being out there in all sorts of weather, every day, taking care of business. Showing up even when you're sick. Dragging yourself out there, even when you don't feel like it or it's inconvenient—that's how you develop responsibility, dependability, resourcefulness, and a can-do spirit.

Who doesn't want those character attributes in their children? Goodness, who doesn't want them in their spouse? Deadbeat, bail-out, no-show spouses and employees are everywhere; somewhere along the line, they missed out on chores. I love chores. They offer routine and rhythm to the day. They're a point of continuity, a constant within the flexing spontaneity of life. Chores have to be done at a similar time every day. You can't fill your life with chores, but chores are like bookends to functionality. If you don't get chores done, you can't start on progress type work.

Chores are like an entrance and exit ramp for the day's activities. Morning chores completed, we can get on with the day; it's like a toll booth: "Have you paid your dues?" At the end of the day, chores give us license to leave the projects occupying our day and come to a safe landing.

Sometimes chores occupy a narrow period of time. For example, on our farm a predator in the pastured chickens creates expanded chores, until we get the troublemaker controlled. Depending on the predator, the answer might be expanded chores like setting up a more protective electric fence perimeter for a couple of weeks until the discouraged predator moves on.

A couple of times, we've had an aggressive and persistent bear get into the feed box on the eggmobile. A bear can rip and tear, making it hard to build something impregnable. Since the eggmobiles are covered in metal siding, we can wait until all the chickens are in for the night, close their door, and then hook up an electric fence energizer to the whole shebang. The chickens don't get shocked because the rubber tires underneath insulate the whole trailer. When the bear comes up and touches the eggmobile anywhere, he gets lit up. Only a couple of nights like this dissuade even the most persistent bear. Dissuading Mr. Bear is a short-term chore.

Most chores are long-term. Milking the cow. Gathering eggs. Spending an hour in the garden each day, preferably early morning before it's hot or late evening after it's cooled down. On our farm, morning chores during the summer are all about moving chicken shelters and feeding the poultry. In the winter, hay feeding for the cows tends to dominate chores. In any case, chores are like a metronome to pace the music of the homestead's projects.

In addition, chores provide a foundation for confidence in bigger things. In the Bible story of David and Goliath, David's confidence to take on the giant came from faithfulness and success in killing a bear and a lion attacking his sheep. On our farm, tackling a predator leads to confidence for tackling giants in life. Outsmarting the predator, and enjoying the satisfaction of protecting the flock, engenders savvy for life.

Some predators, like foxes, raccoons, and possums, respond best to lead poisoning (that's bullets, for the uninitiated). The difference between shooting a gun for real and watching a raccoon die is a world apart from shooting on a video game and watching the dead things resurrect when the game resumes. The finality of death, the writhing of the dying animal, the real flow of blood and finally the closed eyes—this is real life. It's both satisfying because the flock is now safe, and a stark reminder that life isn't a game; it's real, and our actions have consequences. Chores offer that.

Some of my most startling moments have been during chores. The big sow rat that suddenly jumped out of the feed barrel when I opened it. That got my heart racing. How about the time I reached into a nest to gather eggs, and a blacksnake was curled up in there eating an egg? I could see big bulges where it had already eaten two, and the third was halfway into its mouth. I noticed all that after my near heart attack.

Perhaps the scariest thing that ever happened to me during chores involved my daughter, when she was about three years old. She'd come out with me to feed the chicks in the brooder house, which at that time was an old 1950s-style chicken house with a wooden floor about 18 inches off the ground. You had to step up to get into it. She was standing in the door, outside on the ground, and I was up inside feeding the chicks.

Suddenly I heard the unmistakable sound you never forget: the quivering rattle of an unhappy rattlesnake. I didn't see it and wasn't sure where it was, but it was clearly right next to Rachel, my daughter. If my leap could have been recorded, I'm sure it would have been an Olympic record. I literally leaped from inside the brooder and scooped up Rachel in one swoop. The snake was inches from where she'd been standing, and I quickly grabbed a shovel and dispatched the critter. Then I collapsed.

Lest I leave you with the wrong impression about chores being the place for heart attacks, I'll assure you that they're also the place for the most glorious spontaneities. I don't know how many times I've literally stepped on fawns while rolling out electric fence during cow-moving chores. Once I came upon a clutch of wild turkeys and picked a poult up while its mother clucked angrily a few feet away. Special, spontaneous punctuations during chores provide dynamism to routine and help us all to realize how the spectacular breaks in upon the mundane. Watch for it. Enjoy it. Chores are beautiful.

While good character includes the ability to respond to spontaneous needs, to be flexible in a crisis, it also includes a joyful contentment in the routine or even the mundane aspects of life. Without chores the tendency is to drift, to flit from one thing to another and take nothing seriously. Without chores, too, we detach from our dependency on visceral, practical activities. If we have no chores, we easily take for granted the person who washes the clothes, fixes meals, cleans the toilet, sweeps the house. As our ship of life sails along, chores provide an anchor, bearings, and depth readings to keep us on track.

Just for fun, let's itemize a typical homestead list in addition to the household chore list I mentioned above:

1. Feed and water chickens.
2. Open nest boxes.
3. Gather eggs.
4. Take kitchen scraps to chickens.
5. Milk cow.
6. Move cow/sheep/goat.
7. Check pigs—feed/water.
8. Put away eggs.
9. Stoke wood stove.
10. Split/bring in wood.
11. Work in garden.

I purposely didn't include much horticulture work here because plants don't have the daily repetitive time-sensitive requirements of animals. But if you don't visit your garden every day you won't have a successful garden. On our farm, pruning the apple, pear, plum, and mulberry trees is not a chore, but it does have to be done religiously each spring. Ditto grape vines.

All these cyclical, routine activities create a constant in life. None of these homestead chores care a lick about a school shooting in Los Angeles or a drop in the Dow Jones Industrial Average. Faithful daily devotion to the needs of the homestead always yields a return, regardless of the Fed, interest rates, or war in Ukraine. Isn't it nice to know that something responds consistently to attention? In a time when we feel unable to create change in the world's big platforms, our homestead chores yield sustenance as surely as the sun rises in the morning.

Immersing in that alongside our children creates constancy in their lives. No matter what, the chickens need feed. No matter what, the cow needs to be milked. No matter what, the wood stove needs stoking. Many years ago, I had a disc in my back threatening to herniate. One morning when we had to butcher chickens because it was a customer pickup day, I rolled out of bed to start the scalder and couldn't get up. Teresa helped me up, helped me dress, and said, "I'll get you out there; now you have to butcher chickens." Don't think for a minute she said this in a demanding way. It was more, "I'll help you do what we both know you have to do."

She, at that moment, helped me be better than I could be alone. She made me do things I didn't think I could do. It was an endearing moment of encouragement, but what made it necessary was that in a few hours we would have 50 customers arrive expecting plucked, gutted, table-ready chickens. I shuffled out there, and we got it done.

Technically, according to my definition, butchering chickens is not a chore because we don't do it every day, but at that moment, on that morning, that job had to be done and we had to get it done. Someday perhaps I'll write a book with stories about getting it done no matter what. When I think of the fickleness of people today, I realize that these kinds of experiences never dominated their childhoods. Where do you learn to show up on time? To work a full day even when you don't feel a hundred percent? To give more than you get? To see what you can do without being paid for it? To sing when everyone else is frowning?

Chores on the homestead scream into life, "I need you!" How many children grow into their teens without ever having felt needed? Really needed? I see these sociological economic estimates on the cost of raising children these days. The whole tenor is that children are a liability rather than an asset. Viewing children as an economic, if not mental, negative indicates a profound mis-application of youngsters' contributions. We've become such a sophisticated society we've eliminated the need for our children's contributions.

Feeling needed is one of the most primal human requirements. Who needs their children these days? In too many families, children are the ones expressing needs rather than being the ones fulfilling needs. As a result, they don't grow up knowing that things depend on them. They haven't exercised their responsibility muscles and dependability muscles taking care of the chickens or tomatoes. These things don't wait for you to feel in the mood before you care for them. They don't wait for the weather to be nice. They don't wait for your hanging out time to be finished. No, timeliness is everything when dealing with living things. Think about a baby needing a diaper change. You can put it off a little, but only a little. Otherwise, you're going to have an irritable baby and chafed bottom.

One of the most common refrains in any business these days is: "I can't get people to show up." Indeed, the opportunities for folks with good work habits and character to ascend the ladder of success have never been better. The averages are low enough that just a little dependability, just a little can-do, will vault you farther than you ever imagined. If you want kids that vault into success as adults, don't worry about vocation. Don't worry about sports. I don't even think you have to worry about academics. What you need to worry about is character; if you've got that, everything else will fall into place.

And I don't mean the kind of character where people mutter, "he's a real character." I'm talking about the kind of character that shows up early, stays late, works beyond expectations, sees what isn't right and fixes it without being told, and cares about the success of the operation even if that's not in the work contract. Why? Because character is who we are when nobody is looking.

That gets cultivated in the chicken pen. If they don't have water, feed, comfortable shelter, and clean nest boxes, you won't get eggs. Someone will ask: "What's wrong with the chickens?" And you'll have to fess up that you didn't do chores. And you'll realize that if you want scrambled eggs for breakfast, you need to take care of the chickens. You find out the chickens need you as much as you need them. And now it's a mutual harmony you create when you do your chores. When you show up on time. When you watch for problems and fix them before they become crises.

Chickens love their caretakers. They don't care about TikTok. They don't care about Wall Street. They have simple needs but respond abundantly to a bit of conscientious care. Who we become at 15 and 20 is a natural development of who we learned to be at 8 and 9. That's such a basic truth that to even write it seems silly and trite. But folks, our culture lives and acts like somehow the truth is silly. Or that this character development recipe doesn't work.

Our supposedly protective child labor laws don't let teens run cordless drills (power tools) for pay, even though they can sit behind 3,000 pounds of steel and hurtle it 70 miles an hour down the interstate. We shut our youngsters out of adult life and then wonder why they don't or won't grow up. "Why are you stuck in the basement at 20?" The kid responds: "Because I never left." If ever a reason existed to get on a homestead and set up a regimen of chores for everyone in the family, it's the crying need to develop good character traits in our next generation.

Many studies show that old people with pets live longer than old people without pets. The findings point to being and feeling needed. "I've got to get up because the cat needs milk in her bowl." It sounds simple, but such a basic chore has a profound influence on whether an otherwise depressed and giving-it-up older person wants to get up or not. In this light, chores are a privilege rather than drudgery.

Chores beckon with the day's opening call. The wood stove needs you. The chickens need you. The cow needs you. The tomato needs you. Chores set our frame of mind; they establish our reference points for the day. Once they're done, we can do other things; they are the open door to other projects. Not walking through that door until chores are done forces us to be at the door knob every day, every day, every day.

I've often said that chores are my favorite part of the day because I get to make animals happy with my presence. In the garden, I can almost hear the carrots and tomatoes sing in gratitude: "Oh boy oh boy oh boy, he's gonna put a hurtin' on the weeds. Boys and girls, get ready to grow." How many beings do you get to make happy each day?

When I'm dependable, faithful, and responsible, the plants and animals under my stewardship and tutelage literally dance in ecstasy. Their giving to me is in direct ratio to my character. This is why kids raised on homesteads excel in the workplace. It's why they shake hands firmly. It's why they look you in the eye. They're self-affirmed, have learned that when they minister in the arena that needs them and counts on them, the world gives back. Not before the care. Only after the care.

Who's waiting on your kids? Who's waiting on your family to taste and see the character-building qualities that flow naturally from homestead chores? Good character is great for the family, and it's great for our culture. Homesteads are probably the most efficient means of blessing our society with good people.

Chapter 9

Buff: The Homestead Physique

Now that we're through the sociology and the psychology, let's deal with the physical. We're raising a generation of wimps.

As of late 2022, 80 percent of military-age young people in America can't pass the physical to enter the armed forces. Don't assume that I think everyone should enter the military. Don't even think I'm unconditionally extolling the virtues of America's military, especially the way it's being used today. All I'm after here is the shocking reality that in the entire pool of recruit-aged young people, only 20 percent have the physical prowess to handle the rigors of boot camp. That's astounding.

The week I'm writing this the Pentagon has announced it's relaxing the rules of numerous medical conditions in order to widen the pool. These are primarily mental impairments, which, of course, homesteading would cure more than drugs.

The reasons for this failure to qualify are many:
1. Poor diet/food.
2. Lack of exercise.
3. Narcotics.

4. Genetic deterioration.

5. Toxicity—glyphosate, Genetically Modified Organisms, et. al.

6. Stress.

7. Lack of sleep.

8. Dehydration—poor water or lack of enough water.

9. Vaccine side effects.

10. Politics.

Sorry, I couldn't resist. I had to throw that last one in there for fun. If you look at this list, homesteading has a high success rate in solving all of them. For our discussion here, I want to drill down on the lack of exercise part.

When Michelle Obama began her "Let's Move" campaign, she was thinking right. I don't care where your politics are; any initiative that gets young people moving is a good initiative. The amount of time children spend sitting is magnitudes higher than it used to be. Think about how children spent most of their time a couple of generations ago:

1. Tending animals, which meant toting feed, water, cleaning bedding, shoveling.

2. Tending garden, which meant planting, weeding, picking.

3. Gathering firewood, which meant cutting, lifting, splitting, lifting again.

4. Hay making, which meant forking hay or lifting and stacking bales.

5. Digging, which meant using hand tools for trenches, post holes, landscaping.

6. Walking, which meant using your legs instead of a 4-wheeler or truck.

7. Horse care, which meant mucking stalls, hanging harness, currying, saddling, feeding.
8. Culinary duties, which meant handling cast iron, shifting wood, stirring, ladling.
9. Butchery, which meant field dressing animals, skinning, gutting, cutting meat.
10. Spinning, which meant cleaning, carding, weaving, looms, knitting, crocheting.
11. Walking to the outhouse.
12. Toting water from the spring to the house.
13. Laundry, which meant scrubbing on a washboard.
14. Sawing, which meant no circle saws and chainsaws; everything by hand.

No doubt I've skipped some, and if this makes you tired just thinking about it, welcome to the club. I'm definitely not purporting all of this as the good ol' days. But when you lay it out this way, the requirements to move just to exist are truly profound. You couldn't sit around and flip on light switches, turn kitchen faucets, and shop online. Of course, in that day we didn't have the IRS either. That made the time and money relationship a little more blurry.

Remember when Pa Ingalls in the *Little House on the Prairie* series didn't have enough money to buy nails to build a door? He simply joined it with wooden pegs. I'm sure those pegs took longer to whittle than the nails to make, but in those days you could almost live without money. Why? Because you were doing all the things listed above and didn't have an IRS hounding you for a check.

Mechanical advantage is not always the best thing. One summer, when I was home from college, I spent the month of

July hand-digging a subterranean solarium on the side of our house. It's a hole 10 feet wide, 30 feet long, and 5 feet deep. That's a big hole to dig by hand and cart away in a wheelbarrow. I look at that today and wonder how in the world I did it. Back then, I had muscles. Big muscles. You know the ditty "the older I get, the better I was." Ha!

I never participated in athletics, though. How did I get those muscles? Carrying buckets of water. Loading firewood. Stacking bales of hay. A homestead affords all sorts of exercise opportunities to help children develop physically. You don't have to buy weights or invest in a Peloton. Just enjoy your homestead and all will be well.

Many years ago, a nearby family approached us about hosting their 14-year-old son for the summer. They wanted their son to have a summer-on-the-farm experience to turn him into a man. We agreed, and they brought him over each morning and picked him up in the afternoon. I remember well the day he and I were digging a fence post hole. We'd gotten into some difficult rocks, and the ground was dry. The day was blistering hot.

We were dripping sweat and getting after it. We finally got the hole big enough, heaved in the post, and tamped it in. That day this youngster learned how to handle a post hole digger, digging iron, and shovel. It was a good job, well done. What I remember most about it was standing there with him, thinking how tragic that his own dad wasn't enjoying that rite of passage with his son. Instead, I had the privilege of introducing him to Mister Digging Iron.

The thing about physical work is that all of it has a technique. I showed him how to hold the tools, get the maximum effect with least effort, keep the hole sides straight. I impressed on him that you don't get maximum effort by brute force; you pick your attack spot carefully. You look for the seams in the rock and wiggle the digging bar. Technique is everything.

A substantial part of our steward and apprentice training here on the farm is about ergonomics. Loading bales of hay is probably the most consistent and obvious test for whether a young person has learned to use their body. I've graduated to tractor driver today. Looking back from that vantage point, I watch the jerky, upper body struggle as these young people learn how to leverage their body. Often I stop, go back, and show how to bounce the bale on your thigh, using your whole body to set it on the stack.

Shoveling is another big one. If you're moving dirt with a shovel, you want each load to be full. Typically, though, young people not familiar with shoveling chop at the pile of dirt, taking quarter-full shovel loads and expending twice as much energy as necessary. They exhaust themselves quickly and get exasperated that they aren't making much progress. Meanwhile, I'm working half as hard while moving twice as much material. Being able to use your body efficiently makes all the difference in how we view physical labor.

Leveraging your core, using your legs, and learning what you can do is remarkable for encouraging more and better work. At our farm, we move a lot of broiler shelters. A lot. We routinely have smaller-bodied ladies who struggle initially. But learning the proper technique and then building up their muscles through the summer yields accomplished, indomitable young people. One of the most delightful things about the stewardship program is watching youngsters' bodies respond to the outdoor work exercise.

We had one young fellow come from two years in college. He was a smaller built guy and gained 30 pounds in one summer without changing waist size. His shoulders turned into man-bombs. We had another young man who came from college and had a genetically husky build—like me. He went the other way, losing 30 pounds. At one of our seminars, addressing the students, he quipped, "I had no idea my body would respond so magnificently." We all roared with laughter over his choice of words, but it definitely described well what happened.

When I see groups of high schoolers these days, I'm struck by the lack of male physique in the boys. They look like Tyrannosaurus Rex, with anemic little arms. If you want your boys to look like men and your girls to have stamina, a homestead is the best development platform. The human body wasn't designed to look like ET.

We do a lot of school tours here, and generally children under 10 years old have no problem keeping up on a two-mile walk-about. But college students are notoriously out of shape. I've had them nearly pass out or quit halfway through. Many are simply flabby and out of shape. Bringing our bodies back in alignment with at least some of this historically normal movement can go a long way in physical development.

Real life work and activity is the foundation for the Crossfit gym movement. Rather than weight machines, the whole idea of Crossfit is to use your body like it would have been used historically. In the days before front end loaders and mechanical everything, our bodies received a daily workout as an inherent part of existence. Ever notice the physique of folks in grainy old black and white pictures from 1900? My favorite is the lumberjacks sitting on top of a huge log. Ain't no fat on that bunch. Imagine their daily routine.

Who doesn't remember the breakfast Almonzo ate in Laura Ingalls Wilder's *Farmer Boy*? It wasn't Lunchables. It wasn't Danish and Bagels. He must have consumed 8,000 calories a day and stayed tawny, fit, and trim. Something about real life movement makes our bodies supple, firm, and functional.

The relationship between body and mind is well substantiated. Being able to do things, knowing what our bodies are capable of, and feeling good about our physical wellness can go a long way in general mental health. If you don't feel like your body is up to the rigors of survival like toting buckets of water, pushing wheelbarrows, or hiking over a mountain like the von Trappe family in *The Sound of Music*, you'll have a gnawing uncertainty or fear. "Can I get out of a tight spot?" That's a question we should be able to answer with a robust affirmative.

In a context of cultural collapse, the one thing you want is a fully functional body. You want to be able to climb, carry, and run if necessary. Survivors not only have mental toughness like we talked about in the previous chapter; they normally also have physical prowess that enables them to withstand hardship better than others.

My co-author of *BEYOND LABELS*, Dr. Sina McCullough, has an additional take on this physical work aspect. It heats up our body and enables our system to dispel gunk. If you're not familiar with the fourth state of water that occupies our bodies, it's a jell. Known as structured water, when it gets hard it holds toxins, and when it's soft it releases them. Sweating is symptomatic of this liquification of our internal structured water.

One of the best ways to detoxify is through sweating. You can use a sauna, but you can also work up a sweat. A sauna doesn't give you muscles. Working gives you strength and detoxifies—a twofer. And you probably got something done in the meantime. I've never understood walk-a-thons; if you want to do something, why not take contributions for digging multi-flora rose out of a farmer's field? Or digging fence post holes? Or loading hay? Walking around a park for money is a terrible waste of time. Goodness, go pick up trash or plant a garden. Do something productive. Walking around a park isn't productive.

This brings me to my last point on the physical workout—immunological function stemming from time spent in or on the soil. Finland leads the world in connecting the dots between childhood exposure to soil and robust immune systems. In fact, their studies have been conclusive enough to get them talking about how to bring farm soil into town to ensure city kids get exposure. For you entrepreneurs reading this, please consider starting a business that brings farm soil to the city. Perhaps a subscription service where folks would get a permeable doormat (like a welcome mat) bladder stuffed with soil and compost from your farm. You change it out every three months to keep it current. You can stomp on it, roll in it, drag the pets' feet across it, and help everyone stay healthy.

Dr. Joseph Heckman's book *Soils and Human Health* quotes the *English Gardener* from 1699: "spare time in the garden, either digging, or setting out, or weeding: there is no better way to improve your health." He quotes Dr. Benjamin Rush, who in the 1700s told patients "digging in the soil" could cure the mentally ill.

Heckman goes on to write:

"active participation in gardening itself also has a long list of associated benefits. Surveys found that participants in gardening activities felt calmer and more relaxed; felt nature was essential to their wellbeing; and had increased self-esteem, enhanced personal satisfaction, and improved quality of life."

This is a riff off the Hygiene Hypothesis, which says that the immune system is like a muscle; it needs periodic exercise in order to be as robust as possible. We live in such a sterile environment that our immune systems are anemic. When a pathogen comes along, our bodies fall apart due to atrophied immunological exercise. People who know me know that I routinely drink out of the cow trough. This is not a joke. The cows are drinking out of one side and I slurp up water from the other side. Sometimes a little cow snot floats on the water; I try to stay clear of that. But I do this to give my immune system gentle assaults to keep it strong and sharp.

I purposely don't wash my hands much and routinely eat things out of the garden or in the field with dirty hands. Same idea. I could die tomorrow, but so far my sickness track record is as good as anybody I've ever met. I'm not bragging; I'm simply acknowledging a principle that makes sense intuitively and that has now been confirmed in spades in Finland. Want your kids to have robust immune systems? Let 'em get filthy ... often.

Where else is this more doable than on a homestead? Whether you have animals or vegetables or both, you're working in soil, handling manure, rubbing unwashed things. All of that exercises your immune system with a million little assaults;

when the big one comes around, your immunological fortress is ready for battle. Don't fear the barnyard for its bacteria and manure; embrace it for the true immunity builder it is.

A permutation of this idea is wounds. I always have a couple of splinters in various stages of infection. Cuts, abrasions, and bruises come with the homestead territory. Not only do these toughen you up mentally, they also provide additional opportunities to exercise your immune system. Again on purpose, I seldom wear gloves. My hands normally look torn up as a result, but I consider this all part of my wellness insurance plan. I love coming home bleeding and battered from chainsawing multi-flora rose and blackberries as part of a fence cleaning project.

Not only does the clean fence look beautiful after I've been through it, my sweat signifies good detox, and the wounds guarantee an active immune response. Talk about getting a lot done in one project. That's the efficiency and health of a homestead environment. City folks have to plan and then drive to get all these benefits. Unless they have a bed of nails in their rec room. Homesteaders enjoy these benefits without paying for them; we can step outside our back door and enjoy these blessings.

An urban environment seldom affords the chance to get splinters, bleed a bit, enjoy a good healthy bruise, or ingest some dirt. Think about how babies put everything in their mouths. They pick stuff up off the floor and stuff it in. When I see parents visit the farm with a bag full of Handi-Wipes, I cringe. Playing in the dirt, petting animals, and gathering some poopy eggs are all wonderful ways to build immune muscles.

My farmer friends who run apprentice programs confirm a disturbing trend: anemic immune systems in our young people. For example, historically we could teach young people how to gut a chicken and didn't have to worry about bacterial contamination. One farmer told me half of his young people now come down with a bacterial diarrhea response when first exposed to the evisceration table. After a few days, they're fine and finish out the season healthy.

That initial contact, though, overrides their immune system. Farmers who sell raw milk used to offer it without caveats. No longer. Now most of them encourage new customers to take it easy initially. A teaspoon day one. A tablespoon day two. Half a cup day three. They're finding that the natural bacteria in milk that feeds our microbiome can be too much for deprived gut flora. Who knows how much sickness today could be alleviated simply by routine exposure to soil, raw milk, and guts?

Farmers with petting zoos have to place anti-microbial washing stations next to the animals to satisfy liability insurance underwriters paranoid of sickness in visitors. Responsible parents should get their children out to farms routinely to roll in the pasture and inhale wildness. Homesteaders don't have to drive anywhere for the privilege. I'd love to see a study comparing childhood sickness rates between homestead children and city children. I daresay kidsteaders would win.

For a host of physical health reasons, then, the homestead is a great place to hang out. Ah, there's the phrase. Come visit me, and we can hang out on the end of a chainsaw. How about that? We'll be healthier and enjoy caressing our ecological umbilical with purpose and pleasure.

Chapter 10

Welcome to the Real World

We've danced around the video game phenomenon already but now I'm going to dig into it deeper. Imagination is a wonderful thing. Fantasy is a wonderful thing; it's the stuff of dreams, vision, and even mission. You can't do what you can't imagine. Who doesn't want kids with good imaginations?

Before something can become reality, it has to be imagined. But imagination not rooted in reality is foolish. Thinking you could live in a world without gravity, or a world without water, is nonsense. Much more productive is imagining a world with more water, where the principles of nature work better. How about a world with more soil, more earthworms, more plant and animal abundance rather than less?

Imagining how to make reality better is productive; imagining without any moorings to reality is silliness. To not be misunderstood, let me quickly say that I don't have it all figured out. For example, animals can't talk, and yet who would dismiss as unproductive any fairy tale in which animals talk? What? No Aesop's Fables? In these stories, though, we never believe animals can talk. To be sure, I know families who eschew fairy tales out of concern that their children may think goblins and talking goats actually exist.

But historical tales where animals talk or horses fly or a fairy flits to a despondent girl living in ashes (Cinderella) always have strong moral themes. Good literature is mimetic, didactic, and aesthetic. Mimetic means it imitates something real, meaning that it has enough reality in it to not be absurd. Much modern science fiction, books and movies, in my opinion, is not mimetic because it's too absurd to have any grounding in reality.

Compared to Jules Verne's *20,000 Leagues Under the Sea*, for example, modern sci-fi zombies wielding weapons that never run out of ammunition are not mimetic. They're absurd. *Star Wars* is believable, and therefore mimetic. How can you not love Chewbacca? When you're watching *Star Wars*, though, don't you find yourself from time to time realizing, "This is stupid?" Intuitively, if we're grounded at all, we understand when to switch off the unbelievable and switch back on when it becomes believable. *Star Wars* is on the edge, in my opinion.

For this reason, *The Matrix* did nothing for me. It was simply absurd. While I appreciate the overall concept of an alternate reality, or realizing that we're being fed a conspiratorial narrative, the basic storyline is too far out. I can feel the pushback through these pages and don't want to jeopardize my credibility by spending too much time on mimetic art, but I'm laying a foundation for a big concept. Hang in there, please.

In addition to mimetics, good stories need to be didactic. That means they should teach something. A narrative that leaves you blah, or wondering what that was all about, is weak and a waste of time. If you have to struggle to figure out the moral or ethical teaching of the story, it's a waste of time. Good stories should be direct; they should leave you with certain understanding of right and wrong, good and evil, peace and war. If you're confused, it's not good stuff.

Finally, good stories are aesthetic. That means beauty. That means ugly needs to be balanced with pretty. It means poverty must be balanced by wealth—which is not always money. It means wrong is balanced with right. We have a saying about good stories: "the good guys win." Even for all its stories of sin and debauchery, the Bible ends with ultimate judgment on evil, righteousness winning, a beautiful garden, and New Jerusalem.

Violence, evil, and debauchery without redemption is hopeless and helpless. Much of modern art is ugly, and we ought to have the courage to say it's ugly. Chickens in factory farms are ugly. Cows knee-deep in manure in feedlots are ugly. The ugliness of childhood cancer is tolerable because of St. Jude's care and healing. Here at our farm, for several years we had an employee that was a graduate of St. Jude's Children's Research Hospital. Our inner nature yearns to look for the silver lining behind the threatening cloud.

As ugly as COVID was, we fail to see its benefits if we don't recognize that school lockdowns made parents aware of what their kiddos were being taught. It drove people to gardening, canning, prepping, and homesteading. All good stories end in beauty.

For those of you without the privilege of a liberal arts education, I hope this wandering down mimetic, didactic, and aesthetic isn't too off-putting. Consider yourself better educated. Ha! All of this brings us to modern violent and/ or mindless video games, or more broadly, screen time. Even social media thrives on simple shock value, not morality, ethics, or elevating people to achieve more. Too often, it makes us feel inadequate, depressed, and even suicidal. It can be addictive and even more harmful than illicit drugs.

A homestead offers a wonderful alternative life experience to that being practiced by today's modern youth. It offers a strong dose of reality. In addition to teaching questionable morals and ethics, or not teaching much of anything, video games encourage instant gratification and manipulation, both of which create impatience.

When you plant a tomato, you prepare a little peat pot with some soil and shove the seed in, covering it and sprinkling on some water. Then you wait. A child doing this wakes eagerly each morning to run and look at the little peat pot. With wide, querying eyes, the child examines the soil, looking for telltale signs of germination. Usually, a child who plants his own seed already has some background in the project and knows that eventually a tomato will appear. Each morning, the child looks for that little curled green shoot.

Finally, somewhere between day 7 and 10, the telltale clump of dirt above the seed shows signs of movement. A crack is there, and the child runs excitedly to siblings and parents, exclaiming, "It's sprouting! It's sprouting!" By tomorrow, the little shoot appears, shoves the dirt cap aside, and begins to unfurl like a tiny hand reaching for something.

In another week, the first leaves appear along a clear, sturdy stem as the tomato establishes itself as a viable plant. In another couple of weeks, the tomato is several inches tall, sporting robust leaves, and ready to plant in the garden. The parent who gave life to the child now helps the child give life to the tomato; together they take the potted tomato out to the garden, dig a hole, and set the plant in, tamping soil around and watering in well. Every day the plant continues to develop, requiring a pole trellis, some twine support, and soon showing blossoms.

Excitement and expectation build. The child watches, tends, waits, watches, tends, waits, watches, tends, and waits. He sees the honeybees climbing in and around the blossoms. Then the little green orbs as the blossoms fall off. The orbs get bigger. And bigger. It's now been nearly three months since that seed went into the pot on the windowsill. The first streaks of red start to show on the biggest green orb. Not too long now. Under the intense July sun, the tomato turns red in a couple of days and the child, probably under adult supervision, plucks that shiny, succulent fruit and eats it like an apple. Juice runs down his arm. His microbiome dances in ecstasy.

It's been worth the wait. Few things tickle taste buds and satiate like fresh-picked, vine-ripened, garden-grown tomatoes. Compare that to how all of us feel about screens. If we click on something and it doesn't pop up in less than two seconds, we start fussing at the screen. Internet marketing gurus say that the average time spent on websites is about five seconds. Only five seconds! That's our attention span.

Our collective attention spans are contracting toward absurdity. Sound bites are the name of the game. Holding attention is more and more difficult. Church sermons have gone from 45 minutes to 15 minutes—that's how long it takes to sip an espresso. When the cup is finished, we're finished. As a society, we're profoundly impatient.

Think about a video game. Not only does it respond immediately to clicks, it records progress in objective points. The tomato plant doesn't give you a numerical tally each day toward some sort of score. If picking the first ripe tomato is 1,000 points, the plant doesn't tell you on day 45 that you've achieved 500 points. You have to observe the plant carefully, see what stage it's in, look at the weather reports to see if it'll be good for tomatoes or detrimental to tomatoes. All these nuances.

In video games, as with computers, you get the same result from the same sequence of buttons. Every time I hit Delete, I expect something to delete. Everything is predictable and instantaneous. If the gremlins take out my avatar, I click a button and the game gives me a new avatar. If my car wrecks in the electronic race game, the screen gives me a new Ferrari in half a second. No problem.

Folks, this is nothing like reality. We went through the joy of the tomato, but what if that tomato plant succumbs to powdery mildew? Or what if that first tomato has an ugly blemish on the end called blossom end rot? What then? I don't press a button and expect nature to give me a new tomato plant. Nature doesn't fix it or replace it instantaneously. Sometimes I have to wait until next year to try again with a new seed and new plant. Wow, a year to correct my mistake, to learn, to progress. No instant tally marks. No instant replacement. No instant gratification.

Let me ask a question. Is life more like the slow tomato plant, or is life more like the instant video game? Is building a meaningful relationship in marriage more like the tomato or more like the video game? Is commitment more like tending a tomato, or like clicking on a video game? As parents and society struggle with the ramification of this instantaneous everything, I suggest that one of the best remedies is a homestead. Immersing in what is often the slog of the homestead is great to take the edge off our impatience and impertinence.

Just this week, I saw an ad for a lady who offers phone etiquette tutorials for $480 an hour. Have you noticed how young people don't say "Bye" at the end of a phone call? They literally don't know how to use a phone. Conversationalism is nearly obsolete. Or perhaps the attention span is too short to handle a courteous "Bye."

We had a steward who took a leave of absence from Apple to come and spend the summer on the farm. Afterward, he went back to Apple, and when I was out in the bay area speaking at a conference, he gave me a nice tour of Apple headquarters. When I entered the main lobby, off to the side was a receiving room for new hires. It had about 20 cushy chairs and sofas, all occupied by recruits waiting for an orientation session. As I walked through the room, I felt like I was on a different planet. A planet where nobody speaks.

The small room, crammed full of people, was completely silent. Not a word. Everyone was on their iPhone. I felt like I was invading a foreign country. Not a single conversation. It was eerie for an old geezer like me. I've never been in a room with that many young people sitting elbow to elbow, not uttering a word. People aren't even that quiet in church. Screen addiction is not healthy for families or society. To be sure, in our modern world we almost have to spend some time on screens. YouTube certainly offers some enjoyable content. I don't TikTok, Tweet, or Facebook. Perhaps I'm too impatient for social media. Ha!

If you look at political voting preferences in our nation, you see a decided difference between rural and urban areas. I believe much of our nation's partisan animosity and division is a rural-urban divide. What accounts for that? Simply this: country folks have other things to do besides stare at screens. The average American male today between the ages of 25 and 35 spends 20 hours a week playing video games. How many times does one of these folks say, "I don't have time for . . .?"

In rural areas, we drive places, so we actually see the countryside. We watch the rain, sun, clouds, birds, flowers. We watch the seasons unfold and notice if things are dry or wet, cold or warm. We're aware of the branches in the road from the windstorm. We understand the ways of deer, possums, and buzzards. We know we're not in control.

The biggest danger growing up on video games is the notion that the world is at my fingertips. I can control everything. All it takes is a button, and I can make my world do whatever I want. The world is at my whim, at my beck and call. This leads us toward more profound hubris than any of our ancestors imagined. They were tied to feast and famine, blizzard and drought, scarcity and abundance.

In the country and on the homestead, we learn quickly that something bigger than us is ultimately in control. Nature is bigger than me. The urban world shows off the achievements of humans; the rural world shows off the achievements of God. Again, to avoid being misrepresented, let me say quickly I don't think cities are evil or that the world should not have cities. I think cities, like all infrastructure and paradigms, can exceed their point of efficiency. Just like nations can become too big to govern, cities can implode when they exceed ecological carrying capacity.

In a city, we wake every morning immersed sensually in human prowess. Automobiles, roads, utility poles, traffic lights, massive buildings. It's a monument to human cleverness, engineering, and dominance. But immerse yourself in the world of the tomato seed, the garden, the cow, the creek, and walnut tree, and you suddenly feel small. Whenever I walk in the woods, I look for those majestic trees and put my arms around them. Of course, I can't put my arms around the best ones. All I can do is spread eagle out and enjoy hugging something I didn't plant that stands as a monument to God's provision and grace. Yes, I'm a true blue tree hugger.

I didn't choose to put that acorn in the ground (oak trees are my favorite). Probably some mischievous squirrel buried that acorn 150 years ago, forgot about it, or perhaps got eaten by a coyote and didn't live to enjoy his larder. In any case, that acorn sprouted, and through season after season, persisted through the canopy. As it grew, it began emitting auxins to deter competition. Yes, plants fight in the forest. This magnificent oak specimen beat the competition, became dominant in that space, and today feeds my soul with mystery and awe.

On a homestead, we try to do the right things, but inevitably the milk cow gets mastitis. The cucumbers get fusarium wilt. Squash vine borer suddenly reduces our sturdy, robust crooked neck squash plant to wilted nothingness. The neighbor's dog, that cute little spoiled poodle, turns into a ravenous wolf while we're away one day and kills half our backyard chicken flock. We know it was the neighbor's dog because we caught it killing when we drove in the driveway and ran it off. Now what? We have a bunch of dead chickens and a neighbor problem. Good grief.

Unexpected things happen in the city, from power outages to sickness to robbery and car wrecks. And pizza delivery; let's not forget the fun stuff. And good Wi-Fi. But all of that is human-created stuff. We watch people build it; we watch people fix it. We pay utility bills in order to keep the supply coming. People make it all happen, both by supplying and demanding.

But the immersion in something bigger than ourselves, the constant reminder that we're not in control, is ever present on the homestead. Our life is dominated by the magic, spontaneity, and dynamism of cycles, seasons, and schedules we neither create nor control. Urban living is dominated by schedules and infrastructure we built. On the homestead, we always have a nagging sense of being pilgrims and caretakers of a continuum far beyond our comprehension.

In the city, almost everything we encounter has a human-built explanation. Somebody can tell us how to make asphalt. How the light switches work. How traffic lights function. How engines, wheels, subways, and planes work. But who can tell why one seed sprouts and another doesn't? Or how an earthworm makes calcium? Or how actinomycetes communicate? Why did this hen choose to go broody when none of the other 25 did?

This is why every urbanite should do something visceral with life, whether it's growing sprouts in a jar on the windowsill or installing a vermicomposting kit under the kitchen sink. Touching life beyond our design and engineering is important to keep a proper perspective of who we are and why we're here. It's not us. It's bigger than us, and homesteads practically assault us daily with that perspective. It's healthy because it engenders humility.

I don't hold the world on my fingertips like some sort of marionette. This is part of the problem with the climate change and zero carbon movement. First, none of the computer models work going backward; they all freeze the world by about 1400. Until the computer models work backward, I don't put much credence on them working forward. But people filled with video games and hubris think we can click buttons and fabricate answers to things far beyond our comprehension.

The manufactured food movement captures the fantasy of the techno-corporate urban narrative because these folks controlling the story, investments, and government aren't homesteaders. The whole idea that the world would be better off without people living in the country, moving two billion folks to cities, and distributing food out of laboratories is absurd to those of us who plan our day according to the weather, and plant according to the season. But extremely well-educated urbanites living on computers embrace the farmless, farmerless, relocated populace notion as if it's a divine salvation plan for ignorant rural peasants.

What if the money and effort going into climate modeling went instead into practical ecological farming and producing nutrient dense rather than nutrient deficient food? We'd get farther toward sustainability than computer hubris can imagine. The tragedy of the human experience is not that we're lazy; we're not lazy. The tragedy is being busy doing the wrong things. It's being successful at the wrong things. It's misplaced investment.

When you couple hubris with impatience, you have a deadly combination. Rather than humbly realizing things take time, and adaptation takes time, we institute hasty solutions that create more problems than we had in the beginning. In fact, a frequent trajectory in problem- solving is to adopt perceived solutions that create more heartache. We see soil fertility as a problem and offer chemical fertilizer as a solution. We see livestock control as a problem and offer CAFOs. We see cooking as a problem and offer Hot Pockets. We see butter as a problem and offer hydrogenated vegetable oil.

If we wanted to continue this list of solutions worse than problems, we could make an extremely long list. The fact that few people appreciate our own ignorance and lack of control is bad enough. But when coupled with an electronic media-induced impatience cult, we inevitably end up in ever greater quagmires. My dad used to call this "overrunning your headlights." This tendency toward hubris often leads to tyranny via government regulation and manipulation. Adam Smith, the 18th century Scottish economist, countered this tendency with the "invisible hand of the marketplace," pointing out that markets respond and adapt to new information and opportunities. Governments create bureaucracies that plague the country even when they do more harm than good.

Urban populations tend to vote for bigger government because folks in the cities don't appreciate how out of control we are when dealing with the biggest issues of life. We don't control love. We don't control weather. We don't control the cosmos and the rotation of the moon. But in our human-dominated surroundings, we think we can cookie-cutter people, actions, food, and energy like we cookie-cutter roads and subway tunnels.

Staying grounded takes real effort in the city. It doesn't come naturally. To be sure, plenty of country people, like industrial chemical farmers, think they can control everything, but even the most ardent defenders of what is now labeled precision agriculture have a gnawing concern about nature's capacity to thwart man's genius. Why? Because they see the list of antibiotics, for example, that no longer work due to pathogen adaptation. Because they try the latest concoction, and it works for a year or two and then plays out. Because they bought the GMO line about better cotton and lost their crop when the new strain succumbed to a mild frost that never hurt historically normal strains. The GMOs were more fragile.

I'm speaking in broad, general terms. This is why, for example, when something like COVID comes along, the hysteria is far more in cities than in the country. Again, if you look at our nation's response, it was far different in rural areas than urban areas. Why? Because those of us embedded more viscerally in the vicissitudes of nature develop a deep and abiding faith in the power of nature to heal. We watch parched fields turn vibrantly green when the rains come. And rains always come in due time. If we're patient. And we wait. And we don't go berserk.

The final thread on this electronic-everything riff concerns relationships, which I view as extended conversations. If you don't know how to converse, you probably don't know how to build a healthy relationship. The outfit offering dating advice to GenZers developed because many young people don't know how to interact face-to-face; in order to learn this ancient art they pay for tutorials on how to engage in dialogue. They want spouses but don't know how to talk to a prospect. It's a real problem.

While some video games encourage team play, most don't. Game designers know how much easier it is to play independently than coordinating with someone else. If it takes two, you have to wait until it suits both parties. What a horrible complication. Single playing is far superior to doubles. As a result, millions of young people today spend hours playing video games alone. They never have to talk to anyone to be entertained. They never have to talk to anyone to achieve—the game gives them a running tally of their achievement. And they never have to wait on anyone to do the activity; the game is always on the app, ready for play.

A homestead screams, "Team! Partners!" Think about the communal activities of yesteryear, with everything from quilting bees to threshing rings. While we may not duplicate those exactly on a homestead, lots of activities require, or at least are more enjoyable, with two people. Yes, I've built a fence by myself, but it's far more enjoyable with a partner to help hold things and talk as you go along.

The stewards on our farm look forward to chicken processing because it's such a communal activity. The executioner starts what we call the disassembly line. The next person handles the scald tank, picker, pulls off the heads, and cuts off the feet. Several people on a multi-station evisceration table pull guts and hand the birds down the line to Quality Control (QC) before someone at the end handles the chill tanks. The jokes, banter, and serious conversations adorning this communal work make it an eagerly anticipated morning of laughter and learning. The human soul craves this kind of meaningful activity shared with others.

Unlike the mega-processing facilities in centralized industrial outfits, our chicken processing team only does this for about half a day at a time. Then we go do other things. What is communally enjoyable in the smaller home-scale set-up becomes lifelong drudgery in the industrial setup. One protocol accentuates human-ness; the other violates it.

When Hillary Clinton said, "It takes a village," political conservatives castigated her for being unAmerican, not appreciating our rugged individualism. While I don't give her blanket endorsement, she was largely right. And I'm a libertarian for crying out loud. Yes, it does take a village, or at least a family, a tribe, something more than one. Have you met people who didn't have any meaningful relational mentoring? They're sociopaths and weirdos. They struggle throughout life.

Healthy relational partnerships, collaborators on practical human-scale projects, are a natural benefit of homesteading. While my description of butchering chickens on our farm may sound repulsive to some, I invite you to come and be a part of it. It doesn't stink; we honor the chickens and dispatch them efficiently. But interwoven in that process is thoughtful, serious conversation punctuated by some of the funniest and most prolonged joking I can remember.

To take the communal and relational aspect of this one step further, it fills each participant with deep gratitude that the others are there. I can tell you, having processed chickens with just Teresa, just the two of us, it's a lot more enjoyable with a couple of additional people. Too often in our culture we view people as a liability. Goodness, Bill Gates and the World Economic Forum think we need to kill about 75 percent of the world's population to survive as a species. How does that make you feel?

Further, if you really believed that, what could you ethically justify doing? It makes me shudder to think what I could justify if I thought humanity would go extinct unless I exterminated 6 billion people. Talk about hubris. Talk about sociopaths.

Rather than deciding how to kill 6 billion people, how can we figure out how to make people affirmed, happy in their skin, grateful for and toward their neighbors, and build stability and humility into our world? When we look at population data, the one consistent determinant is cultural stability—economically, emotionally, and environmentally. If those three components function well, birth rates drop because parents aren't paranoid about not having enough offspring to care for them in old age.

With that in mind, homesteading offers Bill Gates and company an alternative to stabilizing population growth. Unfortunately, most movers and shakers in the world view homesteaders as inefficient and damaging. They raise cows, don't they? Don't they have animals? Haven't they heard about lab meat and fake protein? Let's eat crickets.

Because homesteads are conducive to "many hands make light work," relationships and team building follow like hand in glove. Few needs are greater in our world right now than a sense of appreciation for our neighbors and family members. Feeling needed comes full circle to feeling appreciated. People who feel appreciated tend to stay away from drugs, prison, and homeless shelters.

Instead of complaining about the Chinese sending fentanyl into our country, why don't we look inward and ask, "What drives people to drug experimentation?" I may be naïve, but I would venture that people who feel needed, affirmed, and purposeful are probably immune to temptations to play with drugs. The cow needs milking. The pigs need slopping. The eggs need gathering. The beans need picking.

Accomplishing all these visceral, meaningful, needed tasks and seeing the cycle from beginning to end—planting to harvesting to preserving to eating—all this establishes a reason to live. Mission, purpose, and personal affirmation. I would venture that if someone kept statistics on drug use that teased out homesteading families from everyone else, you'd see a marked difference in usage. Who has time to talk to drug dealers when the zucchini squash is growing two inches a day?

Our kids didn't prowl around the streets at 2 a.m. They were dead tired and sleeping. We never worried about them coming in at night. They were already in at night. Perhaps we could consider homesteads the best insurance policy to protect us from the things we worry about. Think about how fun your parenting would be if you could eliminate the worry that your kids would have purpose, affirmation, and mission from day one? Imagine if the toddler's greatest ambition was to get big enough to milk the cow or empty the jars from the pressure canner? Imagine if the young child's greatest dream was to throw bales of hay like Dad?

Branded as hillbilly backward places by urban hubris, homesteads offer stability and reality in a world desperate for anchors. For havens. For purpose. Homesteads are not anachronistic, obsolete blights of nostalgia; they are tomorrow's solutions to today's dysfunction. They produce generations of young people grounded in reality and bathed in common sense. That's a good thing for country, critters, and kids.

The bottom line is that homesteads inject reality, humility, and relational function into our families and society. In our modern techno-sophisticated screen-centric existence, we certainly have plenty of distractions assaulting these core family and societal essentials. Nothing derails these distractions as inherently and efficaciously as homesteads. For children to develop into serving-oriented, happy, productive adults, homesteads are the ultimate launch platform.

Chapter 11

More Than Money

Industrial, corporate farms are just as deadly to rural America as oversized cities. The rural exodus began with the Industrial Revolution. Steady factory employment lured millions of farm kids away to bright lights and time clocks. Today, 80 percent of Americans hate their jobs. Most of those are in the city. What did we get in the trade? A lot of soul sucking, I'd say.

This migration to cities left rural towns impoverished on several fronts: economic, social, and technological. I will address all of these, but first let's appreciate the power that comes from population centers. America's founders were beyond wise when they crafted the formula for the U.S. Senate, which was modeled after similar constructs in the states. The one thing they feared more than anything was democracy.

Known as mob rule, pure democracy is a horrible idea. It offers no protection for the minority. If you can get the votes, all is possible. Even destroying the minority. Recognizing that control of real estate is perhaps the fundamental benchmark for a society's stakeholders, the founders adopted two legislative bodies. Congress, the first, was based on a popular vote from

prescribed districts, allocated by population. As the nation's
population grew, the U.S. Congress grew; each Congressional
district contains about 750,000 people.

The Senate, on the other hand, gave states equal
representation regardless of population. This protected less
populated areas from being subjected to the ideas of more
urban areas. People and land were, and still are, a tension. I
remember well, back in the 1990s, I hosted a busload of New
York farmers. Our farm was one of several they visited on a
multi-day excursion. Virtually all of them were losing their
farms due to burdensome taxation foisted upon them by the
cities. Population centers can destroy agriculture by failing to
recognize the disparate cost of governing people versus cows
and trees.

Today, millions of acres of prime New York farmland
sit idle due to this urban attack on farmers. Often legislation
is as devastating as guns and bullets. This is why Founding
Father Patrick Henry said, "Give me liberty or give me death."
William Jennings Bryan said without farms to prop them up,
cities would return to weeds and trees within a generation. But
the cities could afford to exist as long as farmland remained
vibrant. Throughout history, cities have existed as dependents
on the landscape, enabled in size and vibrancy by the robustness
of agriculture. Agriculture is what makes cities possible, not the
other way around.

By and large, cities don't recognize this dependency.
Generally, city mentality holds the urban sector as more
important, as master over the countryside. As hubs of commerce
and power, cities subconsciously and consciously extricate
wealth from the countryside. Urbanites pass food safety laws,
zoning restrictions, building restrictions, and other policies

that directly impede the ability of agriculture to remain viable. Cities tend to suck wealth in all its forms out of the country. Centralized banking, centralized manufacturing, centralized commerce, centralized policy—cities centralize power to the detriment of rural areas.

As our nation moves from a Republic to a Democracy, farmers find themselves increasingly on the minority end of urban perception, which ultimately determines policy. This is evident even in outfits that administer conservation easements. One farm I worked with could not add pastured poultry to its operations because their urban conservation easement holder said the lightweight mobile chicken shelters were new construction, which was prohibited by the easement.

Another easement on another farm precluded building a 200 square foot chicken processing shed; again, the easement precluded new construction. A 200 year old tumble-down barn in a swamp couldn't be demolished and that square footage used in a tradeoff. No, that tumble-down barn in a swamp could only be used where it was. Only city people could be this stupid.

I don't hate cities or the people who live there; I'm just stating the historical record. It's not good, bad, or indifferent; it just is. And as societies become more urban, they naturally dismiss farmers as unimportant. Indeed, America now has almost twice as many people incarcerated in prison as we have farmers. If they could vote (which appears to be coming) prisoners will have more power than farmers. Does that give you pause? Any politician who says we're a democracy, or uses the word democracy to describe America, should be summarily dismissed from office. This is a Republic, which protects the minority, not a Democracy, which destroys the minority.

This urban power amalgamation is reflected in our

common description of rural America as "fly over country."
In other words, there's nothing there. Right. Nothing there
but all the resources necessary to prop up urban scabs on the
landscape. Oh my, did I say that? Yes, I do have problems with
city people who legislate me out of business. But I've written
about that in other places, so I won't go into detail here. The
point is that the flow of money, power, brains, and social fabric
tends to migrate from the countryside and build up in cities
as a culture achieves what humans broadly call success. The
butcher, baker, and candlestick maker are no longer embedded
in villages; they're amassed behind razor wire in urban centers,
and the people who work there pass laws that ensure rural
dwellers can't compete with their brands.

Because I travel widely in rural America, I've had the
privilege of being in places devastated by rural impoverishment.
In Illinois, I spent a couple of days with a farmer who
remembers three high schools in the county. His farm used
to support ten families; today, it scarcely supports him. The
county now has only one school; kids spend nearly 4 hours a
day riding the school bus. Parents can't participate in school
activities because the schools are too far away. The couple of
villages in the county are ghost towns. No gas station, no diner,
no hotel, no hardware store.

Massive farms like his don't buy locally. They buy
supplies by the tractor trailer load from distant venues. No
market for farm production exists locally; it all goes to massive
grain elevators or hog processing plants that squat on the edges
of cities and bring all sorts of social tension to the area. I was
on a ranch in Colorado that, in the early 1900s, supported
enough families to offer an on-ranch school. Other families
from the area attended the school, which fed the children

food grown on the ranch. Gradually those ranches went belly up, consolidated into one massive operation, and the school became an elite prep outfit with dining services complements of Marriott. All the food came from cheap distant places on the industrial junk food truck.

The urban-rural tension and need to balance are not new. In his iconic 1973 book *Small is Beautiful*, E. F. Schumacher wrote:

> *"To restore a proper balance between city and rural life is perhaps the greatest task in front of modern man. It is not simply a matter of raising agricultural yields so as to avoid world hunger. There is no answer to the evils of mass unemployment and mass migration into cities, unless the whole level of rural life can be raised, and this requires the development of an agro-industrial culture, so that each district, each community, can offer a colorful variety of occupations to its members."*

Schumacher also wrote about the loneliness of cities versus the communal vibrancy of the country. You can't disappear in the country. Everybody knows their neighbors. In the country, everyone knows who lives where, what kind of car or truck they drive, what kind of cows they have, and normally even their routine because life is visible. I can see when my neighbors mow their hay, work their calves, or cut a tree. The city offers anonymity, where anyone can get lost and disappear. If you want to hide, don't go to the country; go to the city.

As bustling but anonymous as the city may be, it doesn't offer country hospitality or community. But when the country becomes deserted, it loses social benefits too. A depopulated rural area doesn't offer community. By definition, community requires people. As farms expand and centralize, they disband the communities that created rural wealth long ago.

Imagine my joy today at seeing some rural areas brimming with new life as the homestead tsunami takes hold and thousands of urbanites return to the country. The internet has been a godsend, enabling people to work from home or at least not have to commute to the city office. Forget the commute. Forget the "where do we go for lunch?" Forget the office wardrobe. Instead of a commute, many of these folks can spend their time planting tomatoes or milking a cow. I call that a life upgrade.

One of the most profound benefits the homestead tsunami brings to our culture is what I call an inverted economic flow. Instead of all that money flowing out of rural America, homesteaders reverse that flow and bring money back. They bring urban-acquired nest eggs to invest in property, buildings, livestock, ponds, and equipment like bandsaw mills and chainsaws. If enough of them congregate in an area, the local diner can reopen. And if things continue to develop, the local outdoor theater can be viable again. All sorts of vibrant village life can start anew when money flows back into the rural community.

As things move forward, local businesses spring up, from flower shops to website designers. Many people get stuck on the people part of community, but we must remember that people follow opportunity. Opportunity is often spelled c-o-m-m-e-r-c-e. In other words, business opportunity is what brings people who, in turn, build the community. People don't get together, to live proximate to each other, just for the fun of it. Groups that do that don't normally last long. They collapse.

To partner up we need purpose. We don't partner up

just for fun; we partner up to achieve something we can't do alone. That requires projects, activity, transactions known as trade. Economy is the foundation of functional society. This is why when corruption and cronyism dominate a society, the whole civilization falls into poverty, immorality, and depression. That's why the rule of law is foundational to liberty and opportunity. If you can't trust money and you can't engage in commerce without permission or paying a bribe, everything grinds to a halt.

Imagine what 100 homesteads in a community can mean to the economy. Those folks will need tractors, tools, fuel, chainsaws, wood stoves, supplies like hardware and canvas covers. They'll trade seeds and share equipment, but the aggregate economic power of those households is significant. With that potential customer base, Wi-Fi can come in and even the grid becomes more viable.

As important as economic activity is, social value is just as important. With numerous homesteaders, you have people who can play music, sing, write, cook, call square dances; the social fabric can be rebuilt.

In condominiums, more often than not, people don't even know who lives across the hall. They might nod to each other politely, but everybody lives their lives as individuals. Homesteaders know, or soon learn, that they need their neighbors. Unless you start a commune, work and skill sharing are an integral part of rural living. The main reason is not that we're social beings; the main reason is that in the country, we have much more personal responsibility for life's basic needs.

We have a pump and pressure tank to deliver water from a well, spring, or cistern. We don't have water mains with a crew of engineers and maintenance techs to keep everything functional. Out here in the country, we have to know about pressure gauges, vacuum, low flow shut-offs and all sorts of things just to deliver dependable water. Most of us aren't skilled in all these areas: plumbing, electrical, hydraulics. If we use a cistern, we need to know about masonry if it's concrete, flotation if it's polyethylene.

Ever run a chainsaw? You can cut your leg off in a minute with that thing. You'd better find someone to mentor you in safety and technique before you swing that vicious cutting bar into action. And do you know how much more fun it is to work in the woods with a couple of folks? One guy runs the saw, another couple guys pick up wood; maybe another guy swings the maul to split big pieces, or runs the hydraulic splitter. A group working like this turns an otherwise arduous task into a celebratory shindig.

I'm unfortunately too young to remember hog killin's in the community, but old timers told me as a kid that the families would rotate around Thanksgiving from farm to farm until all the hogs were done for the year. I was fortunate enough to marry into a family that had one of the last family hog killin's in our area. By 1980, when Teresa and I married, the rotation no longer existed, but her family, including cousins, continued to do an all-family hog killin' at Thanksgiving for a couple of years into the 1980s. Some of my fondest memories are of the family camaraderie accompanying those festive occasions.

We'd kill about six hogs early in the morning while the frost was heavy on the grass. The blood on that white frost was quite distinct. After scalding, scraping, and gutting, we'd all feast on a huge potluck smorgasbord for lunch while the carcasses cooled. After lunch, we'd cut them up, cook ponhoss, make cracklins, grind the sausage, and in the fading rays of the setting sun, wash everything down with lye soap to clean the tables, kettles, and utensils. This was social function, homestead style.

Looking back on it now, I know Democrats and Republicans got together for these things. People built bridges instead of barriers because they had to. Sure, you could butcher a hog by yourself, but it wouldn't be much fun. Adding people was a benefit worth setting aside anger and "my way or the highway" kind of thinking. The work proceeded at its own pace. You couldn't cut up until the carcasses cooled. You couldn't pull off the ponhoss until it had cooked several hours. As the matriarchs and patriarchs sampled the ponhoss and gave their directions, the cornmeal and broth finally came together in aromatic and tasty ecstasy.

Everyone would grab a pan, like a meatloaf pan, and stand in line while someone ladled in the ponhoss. One, two, three four. Often we'd fill a couple dozen pans and every family took some home. These activities marked the seasons of the year, the seasons of life, the seasons of provision. They held a community together and built friendships that transcended politics and even religion. When you've worked shoulder-to-shoulder with someone on a project, breaking fellowship over some doctrine or political persuasion is much more difficult.

Another element not to miss in all this is that nobody could hide behind a computer. We were all together, in person, watching each other's expressions, voice inflections, and body language. We bumped into each other. We touched. Old folks held the knife hand of youngsters, guiding them through the butchery process. We smelled each other. Everything was public. Going over behind a bush to have a private discussion wasn't polite. Whispering wasn't polite. Anything that was said had to be appropriate and in the right spirit. Elders set bad attitudes straight quickly. Youngsters desperately wanted to be grown-ups, and to please the uncles and aunts.

Today, when families get together, they generally congregate around some sort of screen, as if that kind of engagement serves social authenticity like the hog killin' I've just described. It can't, of course. As a result, society slips into a never-ending spiral of judgment, prejudice, resentment—all the base elements of dysfunctional social structure. Whether you participate with others in a canning party, garden planting, or processing animals, you learn quickly that other-think is as important as me-think.

You start realizing bridges are as important as walls. One of the most common observations voiced by people who attend a homesteading festival for the first time is the notion that "everybody is so helpful." In a dog-eat-dog world, where intellectual property, litigation, hold harmless agreements, and confidentiality rule the day, homesteaders offer unparalleled openness. Sure, I'll help you bake bread. Sure, I'll help you fix the lawnmower engine. Sure, I'll show you how to pack a bearing.

By nature homesteaders are giving people. We learn quickly how fragile our self-reliance can be. When the water pump freezes up in a cold snap, our independence suddenly isn't very strong. When we're frustrated at the canning jars not sealing, our fierce independence takes a back seat to seeking advice. Every homesteader who comes full of spit and hubris thinking they will build their fortress by themselves, comes to a low point in the first few months. Things don't go as planned. The fence sags. How do they make tight fences? Other homesteaders share their expertise happily, without asking for compensation. It's a different world than corporate America, and one our country could do well to encourage in order to balance out Wall Sreet's often vicious competitiveness.

Finally, homesteading brings creativity and innovation to rural America. I've often mused that if we had on our farmstead the engineering and manufacturing know-how for just one day that goes into building a skyscraper, we could develop all sorts of technologically-advanced gadgets. We could have self-propelled eggmobiles. Video monitoring on cow water troughs. Electric tractors. Electrolysis hydrogen separators on our farm ponds.

One of the greatest impoverishments of rural America is that cities have attracted our greatest minds. Suppose instead of going to space, we'd turn biomass into compost and eliminate both wildfires and chemical fertilizer dependency. In that case, we'd have a much more livable planet and not feel it necessary to pollute another one. Technologies exist to move rural America forward in energy, production, land management, biomass usage. Unfortunately, the best minds seldom focus on agricultural issues, unless they're in industrial corporate empires looking to extract more money out of the countryside. Outfits like Bayer, Cargill, and Tyson come to mind.

Those outfits don't help rural America. They view farmers as colonial serfs, feeding raw materials into their value-added urban channels. I've been encouraged by the creative savvy of numerous new homesteaders I've met, folks who come out of manufacturing or the Information Technology world or even banking. Rural America needs brains. We've been draining our brains to the cities for far too long. When's the last time you heard a high school guidance counselor tell a rising senior honors student, "Mary, wow, your grades are fantastic.

You are one smart girl. Have you ever thought about being a farmer?"

In sociology and anthropology circles, the migration evident over the last few decades is called rural brain drain. My friends in the agriculture community despise my use of the term, but it's true. For a couple of generations now, America has shipped off its best and brightest to the cities. Now some of these folks are realizing the country offers benefits they can't find in a 401(K) plan. It offers benefits that increased pay and another week of vacation can't beat. I, for one, am delighted to see retirees and mid-careerists bringing their urban technology and innovation savvy to the countryside. We need it desperately.

This influx of brain power represents a new wealth stream for rural America. Figuring out a better way to do things on the homestead makes all homesteads more enjoyable places to live. Yes, homesteaders aren't afraid of work, but we're not masochists. We like efficiency too. If we can milk easier, build compost easier, spread it easier, build soil easier, concoct bio-fertilizer recipes easier, we're all for it. Just because homesteaders value different things than other folks doesn't mean we aren't interested in better ways to do something.

The problem with folks who have lived in the country for a long time, like me, is that we tend to get ingrown. We do it this way because we've always done it this way. Many people are stressed that in the next roughly 15 years some 50 percent of American agricultural equity will change hands. With the American farmer now averaging 60 years old, this change in land ownership, equipment, and buildings is going to happen as surely as day follows night.

I've been frustrated all my life at stodgy curmudgeon farmers who won't change. They keep using pesticides because that's what they were taught, or what their dad used. They call me a bioterrorist and Typhoid Mary because I don't vaccinate my cows and let my chickens run around commiserating with red-winged blackbirds. In their minds, it's unspeakable negligence. I say good riddance to these old codgers aging out of farming who tell me I want to starve half the world because I won't accept their belief that farming requires chemicals. Let a new generation of homesteaders take up your properties, bringing their new ideas.

By and large, this new generation of homesteaders did not grow up on farms. They weren't taught the conventional way to do things. I have a friend who graduated from ag college with an animal science degree. He quipped, "I've spent my whole life delearning everything they taught me." The good news for new homesteaders is they don't carry this baggage into their operations. They haven't been told ponds don't work, and you can't put water in a pipe. Nobody told them weeds could feed sheep, and mulch could feed garden soil.

I love the enthusiastic, innovative fresh ideas these homesteaders bring. To be sure, they seldom realize what they will need to learn. But they aren't stunted by conventional thinking and are open to new ideas. In addition, they bring a wealth of new techniques and technology.

If you're a master at something in the urban or corporate arena, don't sell yourself short in the contribution you can bring to a rural community. Every one of them is starving for fresh ideas and new enthusiasm. They're beaten down with prices, pestilence, and policy. You can be the encourager, the spark for renewed rural revival. That kind of wealth is better than money. It's hope.

Maybe you'll be the one that shows me how to install a

biogas digester to eliminate the propane bill. Maybe you'll be the one to build a wood-powered gasification unit to eliminate our fuel bill. Maybe you'll be the one to put a solar-powered electric motor on the shademobile with a remote control. Maybe you're the one who can show us how to develop an affiliate marketing program on the internet. Maybe you're the one who can set up a clever inventory tracking system for our vegetables. Maybe you're the one that can set up our books better to track our financials. I for one am giddy over what our city cousins can add to rural America. Maybe you are that engineer, that software developer, that financial computer guru that can bring desperately needed creativity to the hinterlands.

The homestead tsunami restores rural wealth economically, socially, and technologically. All I can say is, where have you been? And it's about time. I'm ready for gasifiers to run my trucks, self-moving infrastructure, and cheaper construction techniques. Rural folks who want to see flourishing in the countryside embrace homesteaders.

To be clear, that's homesteaders, not folks who buy five acres and spend a day a week mowing it as a magnificent lawn. Not people who build a million-dollar house and a million-dollar horse barn and never plant a garden. Those aren't homesteaders. Yes, they bring some money, but they don't integrate into the rural community like homesteaders. Let's put homesteaders on these small rural acreages. Doing that won't bring the city to the country; that's not what we want. What we want is the best of the city without its disconnectedness to the rural mystique.

That respect and honor toward the country, the people who live there, the opportunities and constraints of the ecology, are what rural America needs. Homesteaders bring it in spades.

Chapter 12

Get Dirty

The themes that grace homestead literature fundamentally differ from the themes that grace mainstream agriculture literature. The differences are quite stark, and one of the most obvious has to do with building soil.

Perhaps part of this is because when you have a little, it becomes more precious. If all you have is a dollar, you're pretty careful about how you spend or invest it. If somebody gives you a pile of money, a dollar isn't as precious and doesn't receive the attention it did before. Most of us are familiar with the work-a-day happy couple that wins the lottery and within a couple of years are penniless, divorced, and depressed.

In many ways, American agriculture won the lottery. Over millennia of soil building, North America offered Europeans seeking freedom and opportunity unequaled treasure. Most archaeologists agree that the soils averaged about 8 percent organic matter. Most of the continent received dependable rainfall, and rivers abounded with fish. The Lewis and Clark expedition found plentiful wildlife, thriving Native American tribes, and deep soils. That soil treasure gave these early migrants a platform to launch a rich civilization.

Early American settlers were bequeathed a soil legacy richer than any other western expansion effort found. Better than Australia. Better than New Zealand. Better than Indonesia. The climate, prairies, Native American stewardship, and multitudinous animals produced a rich and bountiful ecosystem. Modern Americans should remind ourselves that in 1492 North America produced more food than it does today. That is remarkable, and humbling.

While I don't want to downplay the role liberty played in America's trajectory toward wealth, let's not forget that it was also built on a deep soil and resource base that none of our forebears created. In fact, they exploited it.

One of my favorite passages in all my books—and I have quite a few—is from John Taylor's *Arator* published in 1818. John Taylor of Caroline County, Virginia was a friend and contemporary of Thomas Jefferson. Here is a portion of one of his essays:

> *"Let us boldly face the fact. Our country is nearly ruined. We have certainly drawn out of the earth three fourths of the vegetable matter [organic matter] it contained, within reach of the plough. Vegetable matter is its only vehicle for conveying food to us. If we suck our mother to death we must die ourselves.*
>
> *Though she is reduced to a skeleton, let us not despair. She is indulgent, and if we return to the duties revealed by the consequences of their infraction, to be prescribed by God, and demonstrated by the same consequences to comport with our interest, she will yet yield us milk.*
>
> *We must restore to the earth its vegetable matter, before it can restore to us its bountiful crops. In three*

or four years, as well as I remember, the willow drew from the atmosphere, and bestowed two hundred weight of vegetable matter, on two hundred weight of the earth, exclusive of the leaves it shed each year. Had it been cut up and used as a manure, how vastly would it have enriched the two hundred weight of earth it grew on? The fact demonstrates that by the use of vegetables [biomass] we may collect manure from the atmosphere, with a rapidity, and in an abundance, far exceeding that of which we have robbed the earth. And it is a fact of high encouragement; for though it would be our interest, and conducive to our happiness, to retrace our steps, should it even take us two hundred years to recover the state of fertility found here by the first emigrants from Europe; and though religion and patriotism both plead for it, yet there might be found some minds weak or wicked enough, to prefer the murder of the little life left in our lands, to a slow process of resuscitation.

Forbear, oh forbear matricide, not for futurity, not for God's sake, but for your own sake. The labour [sic] necessary to kill the remnant of life in your lands, will suffice to revive them. Employed to kill, it produces want and misery to yourself. Employed to revive, it gives you plenty and happiness. It is a matter of regret to be compelled to rob the liberal mind of the sublime pleasure, for future ages, by demonstrating that the most sordid will do the utmost for gratifying its own appetites, by fertilizing the earth; that the process is not slow, but rapid; the returns not distant, but near; and the gain not small, but great."

In my view, that is one of the most eloquent pieces of writing regarding creation stewardship, written by one of our early patriarchs, as an indictment against our agriculture and a yearning for better management. You can feel the heart and soul of Taylor as he implores our young nation to heal the soil and essentially repent from our rape. This is not the writing of conventional exploitive agriculture. This is the sermon of a soil saint. Unfortunately, the nation did not heed his advice and continued down the destructive path.

That he puts a time table challenge of 200 years on the healing makes it even more dramatic, bringing it to 2018. He looked down through the corridors of time and set a restoration target for the fledgling nation. For him to recognize what agriculture was doing to the soil and then to be that optimistic about its restoration, is truly remarkable. Every nation in every time has its prophets; unfortunately, most populations don't listen. Not then; not now.

This is not the writing of corporate agriculture. It's not the writing of factory farming—I wonder what Taylor would have said about these modern blights. It is the writing of a man who loves his land, individually and nation-wide, and seeks healing touches on it. Homesteaders, as opposed to commercial farmers like me, are blessed with an easier stewardship mentality. They aren't trying to earn a living from their property; they merely seek a life worth living.

Many homesteaders eventually move into the make-a-living space, and I'm their biggest cheerleader. But starting at the homestead level, mentally and practically, offers space in time and thinking to give attention to things that commercial enterprises often miss.

E.F. Schumacher in his iconic *Small is Beautiful* says it this way:

> *In our time, the main danger to the soil, and therewith not only to agriculture but to civilization as a whole, stems from the townsman's determination to apply to agriculture the principles of industry. . . Now, the fundamental principle of agriculture is that it deals with life, that is to say, with living substances. Its products are the results of processes of life and its means of production is the living soil. . . .*
>
> *In other words, there can be no doubt that the fundamental principles of agriculture and of industry, far from being compatible with each other, are in opposition: Real life consists of the tensions produced by the incompatibility of opposites, each of which is needed, and just as life would be meaningless without death, so agriculture would be meaningless without industry. It remains true, however, that agriculture is primary, whereas industry is secondary, which means that human life can continue without industry, whereas it cannot continue without agriculture.*

Schumacher wrote this magnificent thought-provoking book in 1973, during the Vietnam war and the euphoric beginning of the hippie back-to-the-land movement, noting that urban anonymity spawned this earlier city-to-country migration. He noted that it followed inevitably the decades-long migration from country to city. In other words, nature has a way of correcting the balance.

Returning to this emphasis on soil development, then, the homestead community by and large isn't a fan of chemicals, whether herbicides, pesticides, or fertilizer. The homestead community questions animal vaccines and eschews genetically modified organisms. As a group, homesteaders believe in biomass and compost as soil developers. Again, this is not a how-to book, so I won't go into all the techniques and protocols for making compost. You can get whole books explaining compost methodology.

The point I want to make here is that homesteaders, as opposed to urban technocrats and industrial farmers, revere and respect soil. They see dirty snow in Illinois and chafe under the realization that black snow represents national poverty. It's not just something that happens every winter. It's not just happenstance. Real policies and management techniques created it; black snow is nature screaming "Guilty!" at farmers.

Schumacher notes that agriculture has three duties:

1. "to keep man in touch with living nature, of which he is and remains a highly vulnerable part;
2. to humanize and ennoble man's wider habitat; and
3. to bring forth the foodstuffs and other materials which are needed for a becoming life."

He notes that the third, as important as it is, cannot and should not dwarf the first two. The homesteader, due to scale and belief, has a much better grasp of the broader responsibilities toward the land than commercial farming. Without the constraints of earning a living, the homesteader is freer to explore and invest in authentic stewardship, from soil development to wildlife enhancement. Sometimes I wish I could be freed from the business cash imperatives of our farm, depending on some outside source of income that would let me putter and play.

As my pulpit has gotten bigger, an endless number of innovative outfits want me to try their products, hoping I'll endorse them and bring customers their way. I wish I could experiment with every creative person out there. I love new ideas, but running experiments is a hassle. As a for-profit farm-dependent business, our team can only spend a little time on experiments.

These days, one significant experiment per year is all we can handle. Over the years, we've run experiments on biodynamics, immune boosters, livestock supplements, predator controls, soil microbe concoctions, foliar fertilizers, and paramagnetic rock powders. I've probably missed a few; we've been at this a long, long time. From a soil development perspective, about the only thing that consistently gives gratifying results is compost. It's magic.

Most large farms can't conceive of replacing their fertilizer with compost. Fortunately, because we started at a homestead scale and initially didn't depend on the farm for income, we were able to start small. The fact that our initial compost trials were shoveled by hand and only touched a couple of acres didn't matter. Dad and Mom had their off-farm jobs, and I was either in college or initially working at the local newspaper.

The farm's lack of production did not jeopardize our continuing here because, as glorified homesteaders, we subsidized the operation with off-farm income. The downside, of course, was that we didn't have all day every day to be here. But looking back, I'm grateful that we started slow, had some independent income, and had the luxury of what commercial farmers would call "piddling around."

A lot of innovation happens when you're just piddling around. Out from under performance pressure, you can play around with different ideas and techniques. Industrial agriculture and global food corporations disparage homesteaders as failing to contribute significantly to production. But many things eventually adapted to a large scale were birthed at the play-around level.

On a homestead with a little 20-horsepower tractor and rented chipper, in a couple of days' work per year, we can generate enough biomass to create enough compost to cover a few acres. Covering a hundred acres is a different deal. These days, on our farm we've graduated to an industrial chipper, skid loader, and several manure spreaders. We do thousands of tons of compost as our primary fertility engine. But we started with hand shoveling, smaller chippers, and one 1950 ground-driven John Deere manure spreader.

For several years, before we owned a front end loader, we hired a neighbor to come down and load our compost. The progression let us refine our equipment, compost recipe, and application timing. Had we been under the gun from the get-go to pay our salary from the farm, we probably would have made more errors. Or likely would have never even experimented with compost on an embryonic level. It would have seemed too inefficient and too laborious. The initial homestead-scale offered the freedom to start at a doable level and observe results dramatic enough to maintain our commitment at a larger scale.

Rather than being embarrassed about being small, we should be grateful that we have wiggle room. When we have to perform on day one, we spend a lot less time being creative. Smallness gives license to innovate. Never underestimate the power of piddling and puttering. Many of our greatest breakthroughs come when we're playing at work.

I've often wondered why farmers don't take care of their soil. Obviously, when Taylor wrote his book in 1818, farmers already hadn't taken care of their soil for a long time. Hence his diatribe against the nation. Why? Don't farmers know their livelihoods depend on healthy, plentiful soil? You'd think such a notion, such a responsibility, would consume both mind and practice. But it doesn't, and for the most part, never has. The history of civilizations is written in their soils.

One of my treasured books is a 1918 translation of *Roman Farm Management: The Treatises of Cato and Varro*, originally written sometime around 200 B.C. by an unknown author. He noted that long ago "the prairie of Rosea was the nurse of Italy, because if one left his surveying instruments there on the ground overnight they were lost next day in the growth of the grass." In his day, that fertility had already been exploited, and it didn't grow grass like that. The soil stewardship issue has been with us a long, long time.

Just to indulge a bit more on this, consider the footnote by the 1918 translator of this treatise when describing the Italian meadows in a certain area. He noted that monks of the Chiaravalle Abbey near Milan in the 12th century recorded a meadow that yielded 30 tons of hay per acre.

> *"The meadows are mowed six times a year, and the grass is fed green to Swiss cows, which are kept in great numbers for the manufacture of Parmesan cheese. This system of green soiling maintains the fertility of the meadows, while the byproducts of the dairies is the feeding of hogs, which are kept in such quantity that they are today exported as they were in the times of Cato and Varro. There is no region of the earth, unless*

it be Flanders, of which the aspect so rejoices the heart
of a farmer as the Milanese."

Not all farmers rape the earth. Isn't it delightful to read about people who know how to preserve and even upgrade the land on which they farm? I've never seen any place on earth that produced 30 tons of hay per acre. Obviously it's possible. Maybe you'll be the homesteader that brings that kind of abundance and restores that kind of fertility to the property you choose.

I think the primary reason agriculture and soil degradation go together is because the decline happens slowly. If you start with fertile soil, you almost can't do anything to destroy it in a year or two. It takes a long time. The Dust Bowl didn't happen overnight; it took years of plowing before the depleted and naked soil could be picked up by winds. Healthy soil is like chocolate cake; it has glue called glomalin, and a sticky structure.

When things degenerate slowly, we don't see the degeneration. And because farmers see the land every day, they don't mark the changes. Depletion happens incrementally enough that nobody wakes up in the morning and notices a change from yesterday. It's subtle and slow, often taking a generation to change noticeably. By that time, nobody remembers what it was like back in the day; the way things are today become the new normal without shocking and obvious signs of regression.

On our farm, we've done many land development projects, from tree planting to tree removal, building ponds and excavating roads. Amazingly, as soon as the project is done, we almost can't visualize what it was like before we started. As soon as we cut the tree, we can hardly imagine what it was like when it was still standing. A newly filled pond replaces the gullied valley, erasing the memory of the previous eyesore.

Because homesteaders don't view their soil from the seat of a tractor or the chemical analysis from a lab, they know it more intimately through feel, smell, and appearance. It's more like a big garden than a real farm. I've been in many commercial corn fields; the soil is like brick. I don't know how anything grows in it. But a garden tended by a homesteader usually has soil rich in organic matter and wafts the earthy odors of active actinomycetes. A compost pile in the corner is the fuel for the garden; biomass cycling through the compost and onto the garden is what builds that beautiful black chocolate cake-appearing soil.

Homesteaders embrace hand work. Commercial farmers with big equipment tend to become dependent on their equipment. If you can't use a machine, the project isn't worth doing. Little corners and edges? Who cares? Goodness, I'm convicted that on our farm we have enough unkempt edges and corners to feed dozens of families. I'm often convicted about this wasted space and, in moments of practical redemption, tackle them with chainsaw and chipper.

With a smaller acreage, more meticulous care is possible. The eyes-to-acre ratio is different; a small holder isn't overwhelmed by the size of the project. Tackling a quarter acre clean up and development differs greatly from 50 acres of saplings, blackberries and multi-flora rose. I still love hand working in a small area because I can manicure it without heavy equipment tracks. Schumacher was right: small is beautiful.

The flip side of this whole discussion is the small acreage

golf green. Few things are as wasteful as residential estates.
The very name sounds pompous and gentrified. People moving
to the country to show that they can afford acreage without
interaction are a blight on our ecological womb. Ownership
demands stewardship. Stewardship demands bringing
abundance to the community. It starts with capturing all the
solar energy possible, which requires leaf area. Folks who move
to the country just to have a big lawn are not homesteaders in
any sense of the word.

Improving soil quality makes plants grow more
vigorously, with bigger leaves and more of them. Few things
vex my soul as much as seeing someone buy 5 acres, spread
chemical fertilizer on it, then apply herbicides to kill clover
and dandelions, then spend half a day a week on a riding
lawnmower to decapitate the struggling vegetation. This
represents the opposite of homesteading. Unfortunately,
you can drive almost anywhere in America, from country to
suburbs, and pass hundreds of acres in every community that
are assaulted and disrespected this way.

Do you realize people still burn leaves? People rake up
leaves and haul them to the landfill; can you imagine? That's a
travesty against the earth. Homesteaders don't. They have tiny
lawns. In fact, homesteaders cut up lawns into garden beds and
fish ponds. Just enough room for a volleyball net or to pitch a
ball is plenty of lawn; homesteaders understand that lawns are
wasted, unleveraged space. Often homesteaders' whole acreage
is equivalent to the expansive lawn non-homesteaders mow
every week.

Homesteaders know spreading leaves on soil feeds it like
nothing else. Why would anyone burn or throw away leaves?
All leaves should be recycled on site, either directly applied to

gardens or composted and then applied. This kind of attention to detail is the luxury and imperative of small holders because their resource base doesn't afford wasteful practice. Every corner needs to be caressed into abundance. At smaller scale, this is not intimidating; it's actually doable.

Every rural acre falling under the control of a homesteader, therefore, has a much higher chance of being an oasis of soil development than that acre controlled by a commercial farmer or rural townie. Whenever I hear about a property bought by a homesteader, I smile a sigh of relief because the chances are it will enjoy better soil development than it has in a long time.

Homesteaders understand the carbon economy. They appreciate the life, death, decomposition, regeneration cycle. They're not swayed by the mainstream agriculture literature that demeans people who question chemicals. So come on, wanna-be homesteaders. Get some of that soil under your fingernails. Enjoy being a cure instead of a cancer. Grace our countryside with compost piles; make the earthworms dance.

And if you're already here, already building compost piles, here's a slap on the back. Congratulations. You're immersing, embracing, participating. You might not see the kind of dramatic results you'd get with chemicals, but your compost will beat them hands down over the long haul. We're playing a long game here, on our homesteads. Don't forget that. Hang in there. Compost wins.

Chapter 13

Growing Groceries

If one thing defines modern mainstream agriculture, it's an aversion to diversity. Monocultures and single species farming dominate America's farmscape, allegedly the only way we can feed the world.

Farmers introduce themselves by what they produce. Corn and soybean. Dairy. Orchard. Chicken. You never hear a farmer's introduction like this: "I'm a landscape steward and grow a cornucopia of plants and animals to create as many intricate, complex biological relationships as possible, like nature." That would sure set them to eye rolling down at the feeder cattle association annual banquet.

Singularity of production and purpose constrict the average farmer's mentality into a myopic interest and sense of responsibility. If he's a chicken grower, he doesn't care too much about the trees in the boundary fence line. If he's an apple grower, he's not interested in chickens.

Perhaps the most telling proof that so-called modern efficient farming opposes diversity is in its constant war on wildlife. Do you notice how frequently wildlife is blamed for maladies in agriculture? The most common these days is that wild birds bring high pathogen avian influenza to domestic poultry. As I write this in early 2023, we've just seen another 15 million laying hens destroyed and egg prices triple due to another outbreak. Total poultry exterminated in this latest round is 58 million.

If the poultry industry had its way, it would kill every wild bird in existence. Instead of realizing that you can't cram that many birds (chickens and turkeys) in that small a space and have them live and eat out of their own toilet without compromising immune systems, the industry doubles down in their war on birds, both wild and domestic. Blanket pesticides aerially sprayed on crops also kill bees, birds, and fish. Highly acidic manure lagoon slurry kills earthworms on contact.

Traveling through the Midwest is heartbreaking for anyone who understands basic ecology. Ripped-out fencerows attest to the singular crop mentality; gone is the multi-species mixed farm of yesteryear. With cheap fuel, we're not constricted to growing chickens where the feed grows; we can transport that feed to wherever we'd rather grow them. Since people are less spread out across the landscape, concentrated markets in population centers beg for concentrated production in chicken factories. People living on top of each other encourage chickens to live on top of each other.

Biological diversity, earmark of stable ecosystems, doesn't find friends within the mainstream agricultural system. Farmers who have pigs near dairy cows now face criminal penalties for proximate multi-speciation. Produce farmers hate deer; if one enters a field of leafy greens and defecates, scorched earth requirements kick in. Large sections of the field can't be harvested if workers find a turd. This fear, dislike, and battle against wildlife is a benchmark of landscape exploitation.

It permeates the entire mainstream industrial agricultural mentality.

If you want to see an orthodox industrial farmer go apoplectic, start talking about mixing chickens and cows, running ducks under apple trees, and debugging a garden with Indian Runner ducks. Such notions are anathema to the linear reductionist segregated mindset that sees diversity as fundamentally disease-inducing. Fortunately, that tribe isn't too interested in what I'm doing. Members of that mentality don't come around here much. That's a good thing, because it protects them from the trauma of seeing me drink out of the cow trough. Obsessive single-species production infuses mainstream agriculture.

Homesteaders are different. Because their goal is producing groceries instead of commodities, their entire objective is variety. Production is generally about replacing groceries. One of the most common questions in the movement is, "How much can I grow on my land?" Since I'm a livestock guy, when this question is posed to me it is usually about animals. Of course, options differ substantially depending on rainfall, frost dates, soil type, terrain, and location. An acre in Utah or New Mexico will never grow as much grass as an acre in North Carolina.

The same is true with trees. For example, the potential for a one acre woodlot in western Virginia's Appalachians is completely different from the forestal output in Montana. In an effort to squeeze out the most from the least, small holders gravitate toward what permaculture calls stacking. Rather than spreading out linearly with single commodities, homesteaders develop vertically with complementary production. Small holdings inherently require aggressive symbiotic production rather than single-product.

You can't harvest multi-species easily. If you mix cucumbers and corn, for example, you can't mechanically harvest the corn without running over the cucumbers. If you run machinery through the cucumbers, you'll mash the corn. Since commercial agriculture tends toward mechanical harvesting, these kinds of varietal production systems don't work. Birds and herbivores are one of the most symbiotic relationships in nature; birds sanitize behind the herbivores. But if you try to run chickens with beef fattening in a feedlot, the chickens jump into the cattle troughs, tracking in cow manure and then depositing their own in the feeders. Instead of being instruments of sanitation, they are instruments of soiling. But on a pasture, the cattle aren't eating out of feeders, and the chickens can fulfill their role as sanitizers efficiently.

Mainstream commercial agriculture, then, because it's committed to machinery and singularity, finds multi-speciation gets in the way. Rather than being complementary, it's competitive. Add in the paranoia about disease and you have the makings of a solid paradigm opposed to diversity. But due to its scale, the homestead isn't bound by these limitations. Hand movement, hand harvesting, and generally outdoor habitats flourish under species diversity. A homestead's scale and grocery-replacement objective incentivize multi-speciation, which aligns more closely with nature's templates.

On our farm, since we don't lock cattle in a feedlot, we can run chickens with cows any time, and the two species benefit from the other's presence. If you garden by hand, you can easily work around the cucumber vines while picking the sweet corn. And you can work around the sweet corn while picking cucumbers. In fact, the cucumbers can trellis up on

the corn stalks and be even healthier and more productive
than they'd be left to their own devices spreading out linearly
on the ground. You can plant leader-follower in order to get
more production per square foot. For example, early spring
head lettuce, long before it's picked, can be interplanted
with fledgling tomatoes for mid-summer production. As the
tomatoes begin to fade late in the summer, you can interplant
cabbage or beets for fall production. Hand gardening facilitates
extreme intensity.

This is why even the most novice, rudimentary backyard
garden is more productive per square foot than the most
advanced, intensive commercial produce operation. Small
acreage commercial farms like Singing Frogs Farm in Petaluma
or J.M. Fortier in Quebec certainly incorporate innovative
equipment efficiencies, but the other half of their success is
strategic stacking. Eliot Coleman, grand-daddy of the small-
scale produce farm, has invented numerous tools, like baby
leafy greens harvesters run by a cordless drill, to bring quasi-
commercial efficiencies to garden hand work.

John Jeavons, founder and promoter of what he calls
French Intensive or Bio-intensive gardening, wrote the book on
getting more with less. *Square Foot Gardening, The Urban
Gardener* and others bring to the homestead-scale produce
garden a host of production-enhancing ideas. I'm reluctant
to start down this path knowing I'll leave out iconic texts.
Homesteading guru Justin Rhodes' *The Rooted Life* documents
more than gardening, of course, but again you come away from
that book realizing that it's the sheer multiplicity of species,
the complementarity, that moves the production to unbelievable
numbers. I can't name them all, but these wonderful volumes
make backyard gardening not just doable, but also far more

productive than commercial counterparts.

The hand work is what makes it all possible. I remember well spending an afternoon with a zucchini squash picking crew on a large government-certified organic farm in Colorado. Pickers walked slowly through the 20-acre zucchini field behind a massive tractor-pulled in-field packing machine. The machine set the pace. I couldn't believe the number of squashes that didn't get picked. The pickers had to keep up with the machine, so if a squash was a bit too far to grab, it simply got left. I was shocked at the waste. And this was a certified organic operation. When I asked about it, the owner shrugged his shoulders and said that was simply the cost of scale. While the pounds picked per acre were acceptable, the pounds wasted per acre were huge.

It's like round baling hay. In Virginia, the Virginia Tech ag economists say that half the hay ever baled into round bales in the state has never gone through the stomach(s) of a cow. It rots in the field. Round balers, of course, were a hay-making breakthrough technology in the 1970s and enjoyed early adoption out west on irrigated land. But as these labor-saving devices moved east, western dryness didn't move with them. Farmers saved the labor of square bales, but lost half their hay crop to rot due to higher precipitation. Even the non-rotten portion severely deteriorated in the weather and rain, creating nutrient deficiency for the cows and bringing on all sorts of winter maladies.

On a homestead, haymaking is primarily a hand labor affair with little square bales stacked under roof. No waste. I don't know how a farm survives when it loses half its production; imagine any manufacturing outfit losing half of the raw ingredients that come in the back door. Many homesteaders share things like little square balers and even tractors. Spread that infrastructure around several owners, and it suddenly becomes cheaper than the supposedly large-scale efficient equipment of big operations. The key to high production and utilization is the common meticulous care that accompanies hand work over mechanical imprecision.

Even factory farms could greatly reduce pathogens by vacating their houses of the dominant species for a month twice a year and preferably growing something else in there. Like pigs instead of chickens. Or chickens instead of pigs. Or cows instead of chickens. Or chickens instead of cows. But a factory production set-up is not conducive to another species. The financial pressure requires constant flow through. Just like a factory that makes disc brakes can't easily retool to make donuts, a factory chicken house can't easily retool to house cows or indoor square dances.

By their nature, compact small holdings demand diversity, not only for health but for abundance. This includes edge effect and wildlife enhancement. The three great environments of forest, open land, and riparian (water) offer the greatest flora and fauna diversity where two of these environments intersect. Part of manicuring the small holding, then, is purposely creating edge effect. That means meandering field edges, meandering woodlot edges, and strategically located ponds. Not massive lakes for speedboats. But small ponds.

Rather than one big impoundment, a homestead wants multiple swimming-pool-sized ponds. Indeed, water redundancy is a mainstay in small acreages. Large commercial operations tend to centralize and segregate water infrastructure where power or landscape features dictate. But on a small acreage, decentralizing everything is easier and cheaper simply because you don't have to run power lines or water lines as far. A small pond that wouldn't be a drop in the ocean on a big outfit can be a production game changer on a homestead.

Dotting the homestead with numerous small ponds increases the edge effect and therefore the overall stability in the landscape. Using the leeward side of the pond to take advantage of water mass to cool or warm highly localized ambient temperature can move gardening zones one or two tiers. Edge effect can be used where buildings meet the ground, especially on southern aspects where siding reflects sun downward to warm the ground a few feet away from the building.

Cisterns to collect roof runoff might need to be gargantuan to be useful for a factory chicken house requiring 4,000 gallons of water a day. But even a small shed roof 20 ft. X 20 ft. generates 8,000 gallons in a 30-inch rainfall area. That will water 25 laying hens, 2 hogs, a milk cow, and two beef steers a year. This is why I'm jazzed about homesteaders; farmers don't think like this. If I have a herd of 200 cows, they need 730,000 gallons of water a year. I don't know any commercial farmer who would consider building a 500,000 gallon cistern (you don't need the whole year's supply because rains come intermittently to replenish the cistern). Certainly a pond that size is not too big, but a cistern, and the roof to support it, is a massive undertaking. Scale matters. A speedboat is easier to adapt than an aircraft carrier.

For some reason, humans are enamored with big. But I'm enamored with small, because I know it tends to receive more judicious care. Smaller chunks are easier to bite off. The chances of a homesteader guttering a 400 sq. ft. shed roof and putting in a 4,000 gallon polyethylene cistern are much higher than a commercial farmer guttering a half-acre barn and putting in a 500,000 gallon cistern. The homestead-scale cistern is affordable and doable with a backhoe rather than much bigger and more expensive equipment. Backhoes are everywhere. Lots of farmers own backhoes; a homesteader asking for a day of custom backhoe work from a farmer gets the homesteader to a good place and gives the farmer some pocket cash. Both parties win.

Substantial productive improvement is possible with subtle landscape changes. A hand dug swale. A small pond. A dozen grape vines on a steep east-facing slope. Strawberries on the eastern edge of a pond, where westerlies deliver water-tempered air. All the things I'm describing can be done on large acreages, and many have been. But the chance of intricate development like this decreases as the acreage increases. All the projects get big and costly or require government licensing.

At homestead scale, you can scratch around, cut some trees, plant some trees, move fence lines, dig a hole, install a cistern, build a small shed, all under the radar. People are attracted to noise and big changes. Quiet, little changes don't attract attention and can be massaged in without anyone noticing.

Commercial orchards, for example, tend to plant trees in big blocks, regardless of slope and aspect. Air flow and frost pocketing vary greatly based on terrain and aspect direction— does the slope face east, west, north, or south? Large-scale fruit operations, for a number of reasons, generally feel compelled to produce on areas regardless of appropriateness to the fruit. The large-scale producer generally uses less satisfactory areas as well as the most satisfactory areas, up hill and down dale. A homesteader, on the other hand, can walk the land, pick the warm air pockets, frost pockets, dry zones, wet zones, and plant mini-scale on the perfect spots.

When driving through the Midwest in the spring how often do you see large square crop fields with water standing in one corner? Later in the summer, those corners aren't growing anything but a scraggly crop of weeds. A homesteader would cut off that corner and not include it in the cornfield. A homesteader might dig a pond in the middle of that corner to let the water drain in, then plant a water-loving perennial around the edges and use the excess spring water to irrigate in the summer. Or maybe that low ground could be returned to pasture that becomes the favorite succulent grass in the middle of summer when everything else is dried up.

The commercial farmer with massive equipment can't custom-tend that wet corner. Part of his machinery goes through it, of course, but it won't grow the crop he's planting because it's too wet. This kind of inappropriate management happens all over the world in large-scale agriculture. With smaller equipment and even hand tools, the homesteader can leverage these special spots to better use. Unproductive areas that don't fit the single crop plan and otherwise become waste areas can be leveraged to great benefit at homestead scale.

Homesteaders naturally bring this kind of customized development to their properties because management customization is easier the smaller the parcel. I've done a lot of farm consulting over the years, and while I'm awed like anybody else over an extremely large acreage, what I enjoy most is the 5-500 acre property. You can actually see it in a day; you can even walk it in a couple of hours. You can see the boundaries and get a feel for it. That level of understanding yields ongoing benefits when you live there and spend hours getting acquainted.

Farmer and author Wendel Berry talks about being able to love only what you know and being able to know only so much. Homesteaders love every square foot because they can get to know every square foot. Large-scale operations may love it in total, but it's hard to know the nuances of every nook and cranny. Being able to place the right thing onto each square foot at the right time, for the right reason, is the privilege of the homesteader.

In short, I understand my commercial farming friends' antipathy toward "those urban invaders that bring city crime, socialism, and ishy-gishy mentalities." Few things infuriate farmers like a new residential estate neighbor who starts squawking about farm smells and noise. "We were here long before you brought your eyeballs and nostrils," says the farmer. In fact, this became such a problem back in the 1980s that nearly every state passed "Right to Farm" laws.

These laws exempted farmers from nuisance suits about odors, dust, and noise as long as these obnoxious trespasses were committed while carrying out agricultural orthodoxy. I called these laws "Right to Stink up the Neighborhood Laws," but then I've never faced one of these lawsuits and have enjoyed great neighbors all my life. Perhaps I'd see it differently if I'd experienced what some of my farming friends experienced.

Another legal development about this time restricted residences in agricultural zoning. Lobbied and pushed by environmentalists to preserve farmland, and well intended, it met mixed reactions in the farming community. Some farmers were glad to see restrictions on urban invaders. On the flip side, others realized these restrictions prohibited the farmer's ability to plat off pieces of land and supplement the farm income with a small piece of real estate. By requiring only large-acre real estate sales, not only did this lock many would-be smallholders out of the market, it denied land owners the ability to capitalize on a new market—the desire of ex-urbanites to live in the country.

For the record, let me reiterate that I make a great distinction between the residential estate and the homesteader. Both might have 5 acres, but their view of it is completely different. In general, my libertarian tendencies make me side with the market's invisible hand. Let things organically develop the way the market wants. The argument I'll make here is that the view from the country and the stewardship of the land is completely different based on the goals of the newcomer from town. A homesteader who wants to embrace the local agriculture community, trade labor, be mentored, bring skills to neighbor farmers, build compost piles, ponds, and plant gardens and fruit trees—all that is a wonderful thing in the country.

I would venture that, in many areas, hundreds of homesteaders, in aggregate, would bring better landscape stewardship to the country than that same land in the hands of a few large commercial farms. I know that statement will infuriate most of my farming friends, but the truth is that most farmland in America is not cared for well. It's eroding and chemicalizing; a bunch of homesteaders bringing compost and ponds and tended woodlots would be a substantial landscape upgrade.

I have no idea how governmental policy can encourage one and discourage the other. One way is land use taxation, which recognizes the low cost of governance in agriculture. To comply with the tax break, a person must produce a certain amount and sell it; the minimum requirements are extremely low. And normally the acreage has to be at least 5 acres or so. I've seen extremely well run commercial farmettes on less than 5 acres out-produce and out-income 100-acre conventional farms.

Many people lament what they consider carving up the landscape with smaller acreages. But I consider the freedom to acquire a small acreage a cornerstone of independence and the most significant step to disengaging from the cash, corporate, conventional commercial system. I'd much rather see land carved up in little acreages than see thousands of would-be homesteaders denied their freedom to quit patronizing a system they dislike. Lest we forget, this is what drove the initial European flight to America. The promise of "my own place" burned in the hearts of people who had been jerked around by lords, ladies, kings, and queens. Here was an opportunity to acquire property.

Our primal human instinct is to occupy a place, love it, nurture it, and bring it to a successional legacy. That's the same in all kinds of cultures, both eastern and western. In ours, we say you can own it. In others, all they allow is control. But the desire to touch and caress a piece of land touches us in the most primal part of our humaness.

This is a good place to insert another of my favorite passages, written by Benjamin Franklin in 1784 while serving as America's new ambassador to France. Titled "Advice on Coming to America," here are a couple of my favorite excerpts.

> *"The truth is, that tho [sic] there are in that country*
> *few people so miserable as the poor of Europe, there*
> *are also very few that in Europe would be called rich.*
> *It is rather a general happy mediocrity that prevails.*
> *There are few great proprietors of the soil, and few*
> *tenants; most people cultivate their own lands, or*

*follow some handicraft or merchandise; very few [are]
rich enough to live idly upon their rents or incomes; or
to pay the high prices given in Europe, for paintings,
statues, architecture, and the other works of art that are
more curious than useful. . . .*

*Great establishments of manufacture require great
numbers of poor to do the work for small wages; these
poor are to be found in Europe, but will not be found in
America, till in lands are all taken up and cultivated,
and the excess of people who can not get land, want
employment."*

Isn't that great? Good old practical Ben Franklin. What a
guy. In this letter, he captures what is in America's DNA more
than almost anything else: the opportunity to own and care for
property. Most Americans don't appreciate how special, how
rare this is in the world. If, like our family, you've lived in a
foreign country and had your property expatriated in anarchy
and lawlessness, you develop a deep and abiding appreciation
for this unique American value.

Whether it's diversity or hydration or compost, the
homesteader brings both caretaking and resource-leveraging
objectives to the land. Both of those are noble and good.
Resilient land management lives within the heart of the
homesteader as surely as dollars in the heart of the Wall Street
stockbroker. I'll put my money on the homesteader any day.

Homestead livestock menageries and landscape
manipulation are good for land management. They bring soil
and stewardship together in a warm rural embrace, and that's
good for families and our culture.

Chapter 14

Knows Stuff

Have you ever wanted to know how to build a shed? How to hang a door? How to wire an electrical receptacle? How to process a chicken? Build a compost pile? Run a chainsaw? Mill a log into boards? Make sauerkraut? Weld? Fix a small engine?

Homesteaders, over time, learn these skills and a thousand more because our mindset is geared toward self-reliance and saving money. Mastery of numerous practical skills reduces your need for cash. As soon as cash needs diminish, you slow the speed of the paycheck treadmill. A big portion of the disentanglement goal has to do with simply reducing our dependence on cash.

Our culture's modern mentality is to earn enough to pay professionals to do everything for us. Perhaps the most egregious example concerns elder care. The goal is to earn enough money to have a nest egg big enough that we can go into a nursing home at the end. The problem with this objective is that the kids share the same goal. As a result, they're chasing the almighty dollar. More often than not this moves them far away from their parents, breaks up the safety net of the family, and ensures that nobody will be there to take care of Mom and Dad when they need it.

Most homesteaders have almost an opposite-cash objective. Minimalism is a movement that honors living on less of everything. Less cash. Less stuff. Less house. Less energy. Shiny objects, baubles, and accumulation consume the lion's share of the modern person's attention. But what does all that get you? More headaches for your survivors upon your death.

Many people believe that hiring others will free up their time for other things. But more often than not, those other things are simply titillating vacuous activities, like subscribing to celebrity social media accounts. Why would you do that? What's the point? If we don't do something productive with all the time we've saved by hiring professionals to do everything for us, we're not better off than if we had just learned the skill and done it ourselves.

According to business books, people have only three things for which society will pay: an idea, labor, or product. Those are the three things a person can sell to earn income. The author and financial guru Robert Kiyosaki's *Rich Dad, Poor Dad* theme would probably add a fourth: investment. We could debate that, but for this discussion let's stick with the three classic income-earning options. These all add value and involve doing something besides entertainment and recreation.

Homesteaders understand that real-life value comes from participating in an arena of these three options: thinking of something valuable, working at something valuable, or producing something valuable. None of this has to be for money.

What kind of work has intrinsic value? What kind of idea has intrinsic value? Asked another way: If the wheels fall off, what options provide value that will withstand cultural collapse? If you were fighting for survival, who would you want as a partner? Would it be the guy who knows how to write computer code or the guy who knows how to save a green bean seed and plant it the following spring?

Television personality and blue-collar work advocate Mike Rowe leads the charge in honoring what he calls "dirty jobs." He extols the virtues of blue-collar work. This begs the question: Who really keeps society functioning? Is it the banking software expert or the plumber who makes sure your toilet flushes? My intent here is not to demean any work; my goal is to elevate the type of work and skill that our culture generally marginalizes and bring it to a point of honor and respect.

According to sociological surveys, 40 percent of Americans want to work with their hands. In other words, nearly half of us enjoy callouses on our hands and some dirt or grease under our fingernails. But our cultural drift has been to either push those folks into desk work or treat them condescendingly in the public square. When was the last time the school board actively sought blue collar workers for advice? Vocational schools fight for funding, while programs that offer intellectual focus groups receive fat budgets.

Homesteading honors dirty jobs as a viable, valuable contribution to society. I don't know about you, but I've decided that my best security revolves around living proximate to people who know how to grow things, build things, and repair things. If you or your friends don't know how to grow, build, and repair, you're going to be in deep doo-doo if and when the wheels fall off.

Every day we seem to read more disturbing news about

Central Bank Digital Currency (CBDC), which is already well underway in Communist China. Imagine being unable to borrow money because you attend a certain church or gave to a certain unorthodox charity. Imagine being unable to buy a tomato from a farmer whose compost pile put him on the outs with the farm sterility police. Imagine having every transaction tracked and recorded by bean counters and gumshoes.

Perhaps CBDC won't come to America. Perhaps we simply see an economic collapse through hyperinflation. What if we go into more strenuous lockdowns than in 2020 over some new catastrophe? What if store shelves go empty? We live in interesting times in which thinking people realize preparing for disturbance is not paranoia; it's wisdom. Once you come to that conclusion, the obvious next objective involves investing in things that are the least susceptible to confiscation, corruption, or criminalization.

As far as I can tell, what you know and can do can't be taken away by any demagogue, elected or not. So far, nobody has figured out how to stick a straw in your head and suck out your knowledge, including practical skills like welding or packing a wheel bearing. The front line of the economic preparedness crowd pushes gold and silver. But that can be confiscated; it's been done before. Besides, you can't eat gold, and silver won't keep you warm in the winter.

Another darling of the alternate economy crowd is cryptocurrency. Watching the yo-yo values of the various cryptocurrencies makes me wary. In general, I'm not a fan of anything on a computer because it's hackable. Even as I write this book, as soon as I finish a chapter, I print it off so I have a paper copy. I don't even trust the Cloud. I like physical stuff. Paper doesn't have an on and off button, requires no password, and is hard to delete.

If things go crazy, the underground will survive based on ideas, products, and labor. One of the best ideas, for example, will be the person who figures out an efficient barter platform. I'm old enough to remember the barter fairs of the mid-1970s. High inflation, Watergate, and Vietnam combined to fuel a movement to create a cashless society. All part of the hippie counter-culture movement, eliminating cash seemed like a good way to disentangle. But without the internet and social media, these fairs required physical presence.

A chiropractor would set up a table and offer adjustments in exchange for a couple of chickens. Guns always commanded high value. And ammunition. One person would bring eggs and another apples, and they'd trade. Of course, these fairs made the Internal Revenue Service (IRS) start looking at barter, which began showing up on tax forms. These fairs showed the creativity of people to build a completely separate economy. I can't imagine what a resurrection of this would do today with the internet. Instead of physically coming to a place, you could electronically aggregate and create millions of tradable products, services, and ideas. Imagine how less cumbersome barter could be on an electronically aggregated platform?

If the internet has a role in developing a parallel universe, it could connect folks wanting to trade in a no-cash system. Depending on sophistication, you could create value units for various things to keep two parties from having to come to a perfect equitable exchange. If you could store unused chits, it would allow more flexibility. I'm sure greater minds than mine are already working on this, and perhaps it even exists, but I don't know about it. The pertinent question for our discussion on value is, in such a scenario, what would you have to trade?

"I'm the top points getter on Angry Birds. What's that worth?" Can you imagine? I don't think basement video game addicts will bring much value when truly intrinsic worth becomes the standard of exchange. The little devil in me would love to see the day when billionaire gaming executives and social media moguls beg me for a hamburger. Homesteaders inherently bring valuable things to the trading table. "I made a six-inch kitchen knife. I'm selling eggs." Folks, this is real stuff that has real value. It's not some collection of digits in a bank code. No brainiac hacker will take your knife or your eggs without your knowledge. Or confiscate them, like the Canadian government did to truckers' money when they blocked roads during COVID.

A community of homesteaders has valuable trade items of all sorts. Practical skills proliferate in such communities, all sorts of inventions, food, energy, herb and health knowledge, first aid—the list of mastery and product is immense. The resiliency inherent within a proximate group of families like this is immense and beats all the insurance schemes ever imagined.

This brings us to the crux of the discussion. What is more valuable than a pension or retirement portfolio? You could say health. I'd rather be healthy than rich, wouldn't you? If we start with health as a foundational value, what gives us the best chance of being healthy? Eating well, exercising well, living in a soothing place. Sounds like homesteading to me, except the momentary panic when the cows get out. Occasional panic is good exercise for the immune system, kind of like jumping in ice water with the Polar Bear Club. Homesteaders eat well, get plenty of exercise, and generally sleep well without sirens and lights interrupting the night.

Nothing puts you to sleep like a physical project workout. It's far better than exercising at the gym because it's meaningful. You can look back at an accomplishment like a fence or a mow full of hay and know your labor was not wasted. That stimulates satisfaction and fulfillment in the mind and offers a contentment toward sleep. Productive work and exercise are two extremely different things. Imagine if all the effort and calories expended in gyms and fitness centers were invested in real work.

Imagine if city folks donated half a dozen hours per week to weeding and food production. Every 40-acre expressway cloverleaf could grow enough food to feed a hundred families. The city maintenance department wouldn't need to buy any herbicides to control unwanted weeds because folks would chop them out with mattocks. Roadsides could be terraced for garden beds. Land nook and crannies are everywhere and underutilized.

If your work involves growing great food, building compost to feed the earthworms to make better soil to grow more nutritious food, that's a win-win-win-win. Health is not something you buy off the shelf somewhere. It's not a service you sign up for, like a subscription. Health is body, mind, and spirit; homestead living ministers to all. If the wheels fall off, the one thing you don't want to be is sick. You need to be able to move, think, and respond; you don't want to be bedridden during crises.

Beyond health, value comes from fulfilling the basic necessities of life: food, clothing, shelter. If we started building a crash-proof context for ourselves, food is the most critical because it's the one we need sooner than the others. Shelter could become number one if we're caught out in a blizzard. But barring that, food and water are first. How vulnerable is your food supply? Your water supply?

Right now, I know our family has several hundred quarts of canned food. We grew the goodies in the garden and put the jars in the basement during the summer season. We have freezers full of meat. Our neighbor, within walking distance, has the cows where we get our milk. Our chickens lay eggs every day, and our grain for feed comes from nearby farmers. It doesn't come from Ukraine, Russia, or Uzbekistan. We have a personal relationship with these folks.

If the wheels fell off, we'd work out a way to make transactions happen. Of that I'm confident. If all that stored food ran out, we could start butchering chickens, cows, and pigs. Our electric fences run on solar-powered energizers; we don't need the grid to control our livestock. This is simply the way homesteaders live. We never sat down and decided to stockpile food; it's just what people did before grocery stores. Fortunately, both Teresa's family and mine continued this food stockpile heritage.

Our families didn't buy TV dinners. We didn't buy Cheerios. We cultivated relationships with our foodshed rather than relationships at the supermarket. This strategy develops friendships with people who know how to grow, build, and repair versus folks who only know how to push computer buttons. I'm glad I know some folks who know how to push computer buttons, but if things went south, I'm convinced I'll have a better chance of flourishing if I buddy up with the folks who have more hard skills.

Gardening is one of the most basic skills virtually all homesteaders share because food and the well-stocked larder are the frontline strategy to disentangle from the system.

Cutting out that dependency on the supermarket cuts out dependency on a host of things, from cash to distant supply chains. To get a tomato to the supermarket requires a host of smooth-running links in a long chain.

You need the money to make the purchase. That assumes you have a paycheck. That assumes your company isn't downsizing. That assumes the economy is purring along and not jeopardizing the market. All of that is just so you can buy the tomato. Now think about the other side, bringing it to the supermarket. The supermarket needs dependable, loyal, competent employees to put the tomato on the shelf. The supermarket depends on steady affordable electricity to keep the lights on, water to keep the toilets running, propane to run the tow-motor that took the pallet of tomatoes off the delivery truck.

The delivery truck required fuel and another whole litany of people and procedures to keep it running smoothly. Mechanics, fueling stations that in turn depend on energy supply, either petroleum or electric. Roads to run on safely and efficiently. Then we go on back to the farmer somewhere who took a risk to grow a tomato. He needed to get the seeds, labor, fertilizer, whether organic or not. The point here is that the journey to get the tomato to your plate from the grocery store has a completely different tale to tell than the tomato you grew in your backyard garden, picked, and plopped on your plate.

When you start thinking through these comparisons, you begin realizing the value of a tomato grown outside your back door. The value of that tomato sealed in a canning jar sitting in your own pantry, under your own roof, lovingly preserved with your own expertise by your own hands. The homestead tomato shows up to your mouth in a far less risky, less cash-dependent, simpler chain of requirements. I hope if it hasn't been plain before now, the why of homesteading is becoming plainer.

In a society that worships cash, including dividends on investments, I suggest that investing in your garden tomato offers more true value. I'm not suggesting we shouldn't have investment portfolios. But to have a cash-based investment portfolio before we nail down the resilience of our food supply seems like getting the cart before the horse. Leaving one undone while concentrating on the other is simply not wise. Both should be done, but one is more valuable than the other.

The exercise with the tomato could be repeated on virtually everything you eat. Eggs, sausage, apple juice, bread—everything has a story chain and arrives at your plate somehow. The shorter that chain of custody, the less vulnerable it is. And the more skilled you are at growing it, whether tomato or pig, the more intrinsic value you bring to yourself and your household. As you begin defunding that long custody chain, you reduce your cash requirement and create value in yourself. These skills don't develop overnight. They take time; in many cases, quite a bit of time.

Fortunately, clothes are almost a non-issue in these modern times with textile manufacturing as automated as it is. Most of us can go to the thrift store and buy enough clothing to last several years. I have a stash of work clothes that can go a long time. Dress-up clothes are at the bottom of the priority list if things go south. I won't shed too many tears over the demise of the fashion industry.

How about shelter? How vulnerable is your shelter? First, can you be evicted or pushed out? Can you keep warm even if the power goes off? Can you cook? I don't worry too much about heat; you can always take off a few more clothes. Winter is what kills, not summer. Homesteaders often heat with wood harvested from their own land. One of the biggest issues with electricity is energy storage. Even today's most sophisticated batteries require a monthly check and maintenance program.

If you store wood under a roof, it can sit there for 50 years and not deteriorate one iota. That's stable energy. And low maintenance. All you have to do is keep the roof on the protective shed. Wood is the ultimate solar energy. And the smoke induces healthy vapor condensation to maintain consistent cloud formation and rainfall. If you want to heat with something else that's more convenient for now, that's fine, but a wood backup plan is a great value. Wood smoke particulate is completely different than exhaust smoke from burning petroleum or coal. Maverick scientists say we need more smokey fires to give water vapor particles on which to condense to make clouds to create stable weather patterns. Within reason, wood smoke is healthy for the environment. Wood heat can keep you warm and induce stable rain.

Now that we have the necessities of life covered, what's next in our value proposal? I'd say the next is repair. How fast does a handyman project stymie you? I remember well a few years ago when we had a steward applicant from the United Kingdom. She'd quit school at about 15 years old and started working in restaurants. She worked her way up the competency ladder to good pay. Living cheaply, she saved enough money to buy a run-down flat (apartment in the UK, for those who don't know British vernacular).

Using YouTube and mentors, she tore into that flat and remodeled it completely, doing all the work herself. She changed out the bathtub, installed new drywall, put in a new kitchen sink, and fixed the droopy doors. She rented it out for two or three times its original rate and got her investment back quickly. Then she bought another one. By the time she came to our farm for the stewardship checkout, she was living comfortably on the income from several of these fixer-upper flats.

Before she was 30, she had an upper middle-class income from her real estate dynasty, all put together with her own remodeling skills. Self-taught. Just savvy. Isn't that a great story? Listen, folks, if things go south, I want to be friends with people like that. I'll gladly trade a ham or a few quarts of canned squash to have that kind of expert look at my sagging bathroom floor.

How about your car? One of the tragedies of modern technology is that you can't fix anything without computers. How about your chainsaw? What do you know about fixing things? Many years ago, I enrolled in a small engine repair class at the local vocational technical center. I'm not a mechanic, but I've picked up some skill over the years just tinkering. This class was on a whole different level, and it was one of the best investments I ever made.

When we examine the why of homesteading, part of the why is to get off society's treadmill: stop seeking paychecks, recreation, entertainment, status, or social achievement, and start investing in you. I venture that many of the mental and emotional health issues of our day would cease if people started investing in practical skills for themselves. Bringing a lawnmower engine to life is an incredibly satisfying thing.

Taking a wheel apart and packing a squeaky bearing, shoving that grease in there by hand, and putting it back together so it works—that's real affirmation. But when we're scurrying around running to this activity and that activity or commuting two hours a day to the office, we simply don't have time to invest in these personally value-adding ventures. We say we want to improve ourselves but spend the weekend vegging in front of the football game. Or we watch the latest Netflix movie. Why? Because in our hurried, harried lifestyle, we don't have any juice left to invest in our own betterment. The only thing we can muster the energy for is shuffling to the couch and hitting the button on the remote to turn on a screen.

A homesteader's mentality is that these practical skills are not just things we do or learn when we have time; they take priority over other things. Our survival, our personal worth, take a decidedly upbeat turn when we disentangle from society's expectations and invest in ourselves. You know you've entered a homesteader's home when you can't find a TV.

We've grown things, repaired things, and now we want to be able to build things. Whether it's a gasification unit to run a generator for alternative power or a shed to store firewood, homesteads offer endless opportunities to build things. The ultimate, of course, is your own house. Our son Daniel built his own house when he was 20 years old. It's a wonderful house, solid as a rock. We went out to the woods, cut the trees, and milled them on our bandsaw mill. We stickered the lumber and put it in one of the hoop houses we use for chickens in the winter.

Closing the doors on the hoop house created a poor-boy kiln, and we dried that lumber in a few weeks. At 20 years old, how did a farm boy know how to build a house? Since we home schooled, he had time as an early teen to go with work crews and help construct some houses. I was a product of regular public school and did not enter life nearly as skilled in multiple hands-on things as he did. I was on the debate team, though. It's all good. Ha!

The point is he built the house, with some expertise from three neighbors. One knew how to lay block, so he did the basement and foundation block work. Another knew plumbing, and another knew electrical. Of course, Daniel grew up building sheds, eggmobiles, chicken shelters, and other structures here on the farm, giving him a basic understanding. Regardless of what happens to our society, people will always need to build things.

Being able to tackle a project like that without feeling intimidated, and possessing the confidence to muddle through it if necessary, is a natural gift of homesteading. Nothing beats building a structure yourself, standing back, and admiring your handiwork. Homesteaders generally develop mastery in several rural living skills, not just one.

My dad was a master carpenter. He could build anything. Then I came along and I can't make something square if my life depended on it. My tape measures have one-inch increments; nothing else. I never got that half inch, eighth. What's that all about? On our farm, you can tell when structures were built based on the quality of the workmanship. Daniel took over all these construction projects when he was about 15 or 16 years old. They actually look good. Mine look like, well, like it's function over form. Is that fair? I never had a cow say, "I'm not going in that crooked shed." Daniel's buildings look good; mine not so much. But I got them done, at almost no cost, and they're still standing.

The joy of homesteading is that you don't have to satisfy building inspectors. You don't have to pay for construction crews. You can do it at your speed, in your own style, and it reflects you. It reflects where you are in your mastery journey.

The first several structures we built back in the 1970s used recycled lumber from old farm buildings. Dad and I tore down three old collapsing barns in the community to get enough lumber, for free, to build our own structures. They're still standing. We enjoyed seeing how old-time masters connected wood, and we incorporated their techniques into our own construction. Mine looked like parallelograms, but the gates swung and the roofs shed water. That's all that really mattered. It keeps your estate from gaining value and makes for low insurance rates when your buildings look like tumbledown Jack.

I'll finish this theme with two additional points. One is that the residential estate migrants who come to the country but continue to live like they're in the city don't learn these practical skills. They don't build cisterns. If they want an outbuilding, they call an outbuilding company and have it brought in or built on-site. They don't plant gardens. They often hire a lawn care service to maintain the lawn; a landscaper to maintain the flower gardens. The devil on my shoulder sometimes wishes things would collapse so these folks would be forced to live more intentionally. Country living without an interest in rural heritage skills is perhaps one of the biggest cultural wastes plaguing our country. We need homesteaders on these small parcels, not prudes.

Second, successful homesteaders often discover interests and talents they never knew they had. In any activity, people gravitate toward portions that comport best with their passion. On our chicken processing line, which runs most efficiently with about eight people, our stewards and apprentices learn all the positions. Invariably, toward the end of the season, individuals gravitate to certain stations. Everybody CAN do everything, but each station doesn't carry the same allure for each person.

On homesteads, the same thing happens. You find out

if you're more a livestock person or a plant person. You find out if you're more a vegetable garden person or an orchard and bramble fruit person. You find out if you really enjoy mechanic work or if you'd prefer to hammer nails. Everybody who knows me knows I'd run the chainsaw even if I never got paid to do it. As these proclivities develop, your skill level will increase in that activity. Mastery often develops into a side hustle. And sometimes that side hustle can become a full-time salary, perhaps the Holy Grail of homesteading.

When most homesteaders start they never imagine their place could replace their paycheck. But along the way of doing and tinkering and learning, they discover abilities their school guidance counselors and academic family members never encouraged. In five years a homesteader may find she's really good at making sauerkraut and start a business. Practical skills never disappoint. Who's disappointed that they know how to do something? We're only disappointed at the things we don't know how to do. Be careful; homesteading may lead you serendipitously to your life's greatest vocation, and it may look entirely different than what's bringing home the bacon right now. Homesteads offer these kinds of discoveries because when we're doing productive, meaningful things, we develop. That's the way life is.

I hope this discussion opens your eyes to investment value beyond cash and cash portfolios. Cash helps us do things, but it can't affirm resiliency and personal practical skills. The more eclectic mastery we have in growing, repairing, and building, the more value we have in ourselves, intrinsically, and therefore the more value we bring to our families and our communities. Everybody needs somebody who knows how to do stuff. Being that person is the most stable retirement plan of all because it endears you to lots of needy people. Serving our own needs and the needs of others is the best fallback plan we can have. Practical skills from homesteading give us a return on investment that is authentic and functional.

Who doesn't love the guy or gal that knows how to do things? Every group or organization has that person who seems to know how to put things together and make things happen. Homesteads cultivate those kinds of folks. I'll bet I have plenty of readers right now daydreaming about becoming that person. Go ahead. Imagine it. Your homestead may just be the ticket to being everyone's favorite team member. Just like the super jocks on the school playground, you'll be the first pick. Successful homesteaders often become the most wanted simply because they know how to do stuff.

Chapter 15

Pretty Panorama

The nature beauty business is big. From national parks to dude ranches and safaris, the personal desire, need, and investment surrounding immersion in natural beauty is nearly unquantifiable. How far do you have to go, how much do you have to invest, to get away and commune with nature's beauty?

Imagine if you could enjoy it in your home and yard? If stepping off the back porch immersed you immediately in the magnificence and majesty of nature? Here on our farm, every sunset and sunrise is breathtaking; some more than others, but all are captivating and bookend the day with beauty. What's that worth?

Unfortunately, farms have historically not been considered places of beauty. If we think back over our country's history, we don't see many farmers writing about the beauty of their place. Farmers write about how to produce more or grow things faster and easier. Farmers typically don't write about the inherent and innate beauty of their places.

Modern conventional industrial agriculture prejudices the average city person with an aversion to farms as dirty, ugly, dusty, unwholesome places. Indeed, the average farmer doesn't live as long as their city cousins, probably due to exposure to chemicals these days. In any case, in our culture, farms are not considered repositories of beauty; they're considered production units for raw materials. Often they carry a stigma of stench and unsightliness.

Who wants to visit and walk through a CAFO chicken house? Who wants to visit and walk through a sterile field of corn? While these kinds of production units carry an aura of awe due to their sheer magnitude, they don't overwhelm with innate and intricate beauty. They're monuments to hubris, not mementos of God's handiwork.

As a result, many if not most urban folks have an aversion to visiting farms. Why would they want to come and get stinky and dirty? This is the average assumption. Perhaps the most common remark I hear from visitors to our farm is, "Nothing stinks, and it's all beautiful." They're shocked, surprised, and pleasantly informed that farms don't have to be cesspools of sensual obnoxiousness. Remember, good farms should be aesthetically and aromatically sensually romantic.

Because a homestead is not pushing production every minute on every acre, folks enjoy the opportunity to value natural beauty. While homesteads can mismanage and turn ugly and stinky, most small-scale operations create landscape attractiveness easier than large commercial operations. Things happen at scale, which makes bad things worse.

A couple of pigs on deep bedding in a shed is one thing. A hundred pigs on deep bedding in a shed is quite another. In your home kitchen, keeping things clean isn't as difficult as keeping a commercial restaurant kitchen clean. At homestead scale, building in beauty is not only logistically easier, it's also mentally easier. Because homesteaders can focus on things besides performance, they can pay a little more attention to adornment.

Homesteaders routinely develop their properties like a jeweler cuts a diamond. Because nature is the template, here are some things homesteaders employ that enhance the landscape's primal beauty.

HYDRATION

Understanding the role of beavers in prehistoric landscapes requires us to appreciate that in 1492 some 200 million beavers inhabited North America. Creating a landscape of 8 percent water, these beavers were perhaps the first massive casualty of European colonization. Prior to synthetic fabrics, wool and beavers provided the lion's share of jackets.

Cotton was expensive and not warm. Beavers were easy to trap because they made trails where they came out of their ponds seeking new trees. These well-traveled paths concentrated their movement into funnels that made trapping far more efficient than trying to trap a fox or even a mink.

Most homesteaders realize the value of a pond and build one soon after arriving on the property. From a home entertainment perspective, nothing beats a pond. What child can't spend hours casting a line and lure, hoping to land a fish? The catch doesn't have to be big. Even a tiny bluegill elicits screams of pleasure. People pay big bucks to go places to fish. The average homesteader can enjoy fishing a few yards from the house.

A pond offers constant wonder, from the turtles that surface like little buoys to massive mud turtles floating, suspended between surface and depths. What child doesn't startle with glee at the sudden ker-splash of a frog jumping from its water-line perch into the safety of the pond's deeper waters? In late evening, stealthy sleuthing along the bank often yields close encounters with massive bullfrogs. The deep-throated harrumph of these pond denizens calls for exploration. Being able to spot one before its fright hurls it into the pond is a game you can play with your kids for hours.

Sitting on the bank on a summer afternoon counting dragonflies scurrying back and forth is as entertaining as anything I've ever done. Their aerobatic ability to flit, hover and turn is truly amazing. Where do you see that in the city? Soon, a raucous red-winged blackbird lands on a cattail nearby, squawking to scare you away from her nest of young 'uns. Surrounded by such a dramatic, active theater is like sitting in the midst of the best auditorium in the world.

As things cool off in the afternoon, you might see a muskrat waddle out of the water and take a mouthful of grass off the bank before slithering back into the water, swimming to a hole at the water's edge. You can't see inside the hole, but you can be sure some little ones will soon be born in a grass-lined nest. Once in awhile you'll have the privilege of seeing a deer come gently and stealthily to the pond for a drink.

One morning, walking up quietly to move chicken shelters, I heard a commotion at the edge of the pond and approached extra quietly to see what it was. There were two bucks going at it, antlers entangled then apart, then crashing and thrashing. About 20 yards off to the side stood a watching doe. Oblivious to my presence, the bucks continued charging at each other as I stood transfixed for several minutes. Finally, one buck clearly overpowered the other and ran off with the doe. Lots of people would pay big bucks (sorry, couldn't help the pun) to see something like that. On a homestead, these things weave into the routine as naturally as stars twinkle in the night.

I remember one winter when we were ice skating on one of our ponds, a deer came down the bank, stepping onto the ice. Its legs went every way but straight. We were on the other edge, putting our ice skates on. The deer hadn't noticed us when it approached the pond. But once it started losing its legs in a jumble of twists, we laughed and startled the deer. It turned its attention from trying to stand upright to trying to escape. With all its legs splayed out, lying on its belly, the poor thing finally got enough traction to hump itself over to the edge of the pond. As soon as the front legs hit the bank, it scampered off into the woods like an arrow shot out of a bow.

That happened fifty years ago, but I still remember it vividly and fondly. Those are the kinds of things you see around a pond on a homestead. I've picked up baby turkeys, stepped on fawns, and even stepped on a rattlesnake. You can't plan these things. They happened to me because I was present. If you put in enough time, these kinds of experiences will adorn your memory. You don't have to travel, leave your bed, or leave your house. It's a perk of the homesteading experience. I'll put it up against screaming fire engines, car wrecks, and drive-by shootings any day. They might be memorable, but they aren't as beautiful.

BIOMASS REGENERATION

Can anything be more beautiful than a compost pile? Perhaps nothing illustrates ecological assault better than putting biomass in landfills. Roughly 80 percent of everything put in landfills is compostable biomass. Disrespecting the decomposition foundation of regeneration is immoral and unconscionable, but this assault continues today.

If one object defines a homestead versus just a home, it would have to be a compost pile. Building a compost pile is the ultimate monument indicating that a person understands ecology, cares about stewardship, and enjoys participating in caretaking responsibilities. Those are noble and sacred life themes.

Think about all the ecology projects people travel around the world to do. Large environmental organizations burn up a lot of computer power, meeting time, and travel energy getting people to an area to do good. I'm not suggesting this isn't important. But with a homestead compost pile, you can do good any time, every day, without burning an ounce of gasoline or using internet computing energy.

On a homestead, compost is the backbone of the garden fertility program. You can see the results throughout the growing season, and it's nothing short of miraculous. Any biomass generated on the homestead, from manure to leaves to junk hay and garden weeds, feeds the compost pile and, after digestion by microbes and worms, yields a feast for plant roots. Isn't it wonderful that God set all this up in a way that offers each acre of land its own renewable path? Fertility doesn't require expensive inputs from across the globe.

Everything a square yard of soil needs can be generated right there. Worms, minerals, microbes, water, and air find sustenance on site. Talk about divine provision. When a homesteader spies some half-rotten round bales at the edge of a neighbor's field, he knows that's more precious than gold. The neighbor will probably sell it for a song to get rid of it and the homesteader turns that wasted biomass into humus for the soil. Black, rich, earthy-smelling soil is beautiful like nothing else.

Compost is the catalyst.

Rather than locking land up in an abandonment scheme like wilderness areas, homesteaders see opportunity in working with the land. Participatory environmentalism using our intellect and mechanical ability to come alongside nature as a benefactor rather than a spoiler—that's the attitude and dream of homesteaders. Humans can touch land in horrendous ways, that's true. But we can also touch it in beautifully healing ways, and I contend that not only do homesteaders want to caress the land, they also get to enjoy the beauty their touching creates.

Homesteaders enjoy the progress from year to year. When our family came to this land in 1961, the garden just north of the house had been chemically fertilized and tilled excruciatingly for years. The red clay soil was like bricks. Clods were as hard as rocks, and of course we kids would throw them at each other from time to time. Our predecessors on this land did not share our stewardship ethic or our understanding of biomass decomposition.

We began applying copious amounts of mulch. Lawn clippings, leaves from the city's dump pile, and old junky, rained-on hay supplied many inches of mulch. Within a couple of years, things began to improve. Within a decade we had earthy, black, fertile soil. Today you can stick your hand down a foot; the soil is loose and looks like chocolate cake. Smells delicious too. In my lifetime, to watch that hard-scrabble soil turn into rich black earth has been one of the most beautiful and gratifying things to witness.

Anytime I want to confirm this success, I can step out the back door and take a look. I don't have to go somewhere else. I don't have to get in the car, on an airplane, show my passport, or hassle with travel restrictions. I just step out the back door and feast. That's true home entertainment.

PREDATORS, PERENNIALS, AND PRUNING

The pristine natural places people pay to visit usually teem with activity, although it's not human activity. Imagine what things looked like when dinosaurs roamed the earth. Watch any nature video, and it's full of animals, insects, and birds. Predators chase prey and eat in a bloody, gory, butchery feast. Some may think that's not pretty, but the lion would disagree. It's primal nature, which is highly competitive, active, and diverse.

Homesteaders constantly deal with predators. The favorite prey, of course, is chickens. Everything that goes bump in the night likes chicken. You may wonder why I would include a raccoon decapitating a couple of chickens as something beautiful. What's beautiful is catching the predator and otherwise thwarting the little devil. Video games and athletic games can't begin to compare to the adrenaline rush and stakes in a real-life homestead predator match.

First, you have to identify what it is. Skunk, raccoon, fox, dog, cat, mink, weasel, bear, coyote, wolf, hawk, pine martin, owl—all have telltale operational habits. Learning their signs and then outsmarting them is a beautiful thing. Having battled predators all my life, few things bring me a deeper satisfaction than terminating a perpetual predator pickpocket.

One year we had a rat invasion in the chick brooder house. I couldn't get ahead of them in the moment, so I moved a cot out there and slept with the chicks. My presence kept the rats at bay. I didn't kill the rats at that time, but I got those chicks safely out to the field where rats don't go. I dealt with the rats once the chicks were gone. Giving that level of care in a vulnerable situation offers its own kind of beauty, and homesteaders exult in this kind of joy in spades.

Most plants are perennials, not annuals. Agriculture revolves around annuals, but nature accentuates perennials. Homesteads tend to exhibit this kind of balance too. Most vegetables are annuals, but the pastures, orchards, and bramble fruits are all perennials. Homesteaders don't build feedlots and CAFOs; they generally build their livestock operations around perennial pastures. This is nature's way, and it's beautiful.

The sheer diversity abundance in a perennial pasture offers a tapestry of blossoms, spider webs, grasshoppers, and bees. Only God could mix and match colors, designs, and sounds in such a three-dimensional display. It all fits and offers pleasantness to the senses.

One of my favorite activities is to go out right before dark on a summer evening and lie down in the pasture where I moved the cows in to graze a few hours earlier. Naturally curious, the cows come over, timid and gentle, to check out this still body. I close my eyes and listen to them sniffing, sticking their tongues up their noses, snorting, and chewing grass. Eventually, the most adventuresome puts her nose on the toe of my shoe. She sniffs heavily and then moves on up my legs. She licks me along the way with that sandpaper tongue, tasting salty sweat.

A second cow joins the first and before long, I'm surrounded by these beautiful placid beasts, sniffing, investigating, nose-rubbing, nibbling my hat, and licking my ears. I've never been stepped on, even when doing this in large herds. They're gentle giants, checking out their benefactor and friend in a marvelous show of respect and interest. What could possibly be more beautiful than to open your eyes and see these contented, sweet cows looking and licking as if to ask, "Don't you have more important things to do?" My answer: "No. You're too beautiful."

I've never been to a Disney park, but I can't imagine anything as magical as being slobbered over by massive cows. Remember, I'm on the ground and looking up at 1,000-pound-plus hooved beasts that could kill me in an instant with a well-placed kick or head butt. But every time I do this, I can almost sense the cows' understanding our benevolent relationship. I place myself at their mercy, and in gratitude, they acknowledge and honor my presence. I'll bet if more people participated in this display of mutual respect, we wouldn't have as many people joining the Fake Meat Demon Cow Cult. Maybe I can take Bill Gates out on one of these reconnaissance missions someday.

Or think about pigs, the most intelligent of all domestic livestock. And the most human. The difference between pigs and cows, of course, is that pigs are omnivores. It does give me pause to realize that when I sit with them, I have to constantly protect my fingers, ears, and nose because they have no reservations about making me their next meal. The mafia have hog farms for a reason.

Like all other groups of animals, a group of pigs contains both intrepid and timid individuals. They all have different personalities. One is the most affable and sidles up with that telltale heavy snuffing. Pigs smell better than they see; they get a lot of their awareness from their nose rather than their eyes. They lead with their noses, sniffing and snuffing heavily. Then they nibble, and their teeth are sharp. Cows nibble, but they do it with their lips, and they don't have front teeth on their uppers. Pigs have sharp teeth front and center, upper and lower.

The affable pig, though, quickly responds to my touch.

Some like to be scratched on their foreheads, like a cat. Others' favorite scratch spot is right above the tail. And yet others enjoy a good belly rub. Those are my favorites. As soon as I begin rubbing the belly, the pig leans into me, as if to say, "Boy oh boy oh boy, that feels good." If the pig is especially responsive, it'll soon flop clear over on its side, as if in an ecstatic trance. Eyes closed, legs stretched out, those pigs will often let you rub their bellies for many minutes if you have the time and don't tire.

What could be more beautiful than a comatose pig in a belly-rub-induced porcine hypnotic trance? I don't have to get in a car or airplane or go through airport security to enjoy this level of beautiful entertainment. It's as good as opera, Broadway, soccer games, and a movie theater. When you think about what people pay for entertainment, homesteaders have them beaten by a long shot. A homesteader doesn't have to schedule any of these experiences; they're there for the enjoying, anytime, and every time.

Where was I? Oh, perennials. What can be more sensually satisfying than a field of fresh-mown hay? Or a mow of freshly stacked hay? Or cows eating hay in a shed during a blizzard? One of my favorite wintertime sights is to go out to our hay feeding shed during a heavy snow and watch the dry, warm, contented cows all lying down on the carbonaceous diaper (if you don't know what that is, read my book *POLYFACE MICRO*) chewing their cud. How can I be mad at the world when the shed I built to protect from the vicious elements houses cozy cows who have not a care in the world? They don't care a hoot about taxes, wars, crime, or famine. All they know is I'm taking care of them, and they don't have to think about anything.

Grass consumption is all part of the pruning that happens in nature. Giraffes, elephants, and birds prune and peck vegetation to freshen it and convert that solar energy into urine and dung, muscle, and sinew. Digestion is like a warp speed compost pile. What a compost pile might do in half a year, an animal's digestion does in a day. Isn't that a beautiful thing? And entertaining. The leavings of the prunings offer endless ventures into discovery and entertainment. You can tell a lot about a cow's health by studying her poop. Around our house, nearly every dinner conversation includes at least one foray into the magical world of poop.

One of my more enjoyable children's activities is taking a bunch of 6-10 year-olds out to the field with magnifying glasses and popsicle sticks. I find a fairly fresh cow pie, something 5-10 days old, and have the kids start rooting through it with their popsicle sticks. Of course, some think it's yucky and hold back. The intrepid explorers jump right in and quickly jabber about everything they find. Dung beetles, fly maggots, gnats, tunnels—cow pies are amazingly entertaining. Again, homesteaders don't have to drive anywhere, fly anywhere, or spend the night in a hotel to be entertained by cow pies.

Did you know that wildlife biologists routinely count deer per square mile based on pruning aggression they observe in vegetation? The higher the population, the harder it prunes the vegetation. As the deer get more numerous and food becomes more scarce, the most palatable woody species show signs of heavier grazing. How far down the twig a deer grazed is directly proportional to the population density in the area. Any homesteader can observe this kind of drama in their own woodlot.

The best homesteaders, who move their livestock rapidly

from paddock to paddock, are pruning aficionados. Like the ancient bison grazing across the prairies, freshening the biomass with pruning, the homesteader's livestock stimulates biomass production, soil building, and ecological vibrancy. A good homestead is a microcosm of heritage ecological systems. Incorporating water, biomass regeneration, and animal activity, homesteads tap into nature's most ancient success patterns, creating beauty and offering amazing entertainment at the same time.

In the next chapter, we'll examine that entertainment and enjoyment part more deeply. Homesteads are beautiful and productive places. Nobody on a homestead should feel embarrassed or apologetic because, "I only have four acres." The truth is those four acres have the capacity to be more productive and more beautiful than any four acres in a mono-species industrial farm. Pick your spot, and enjoy the beauty you've helped create. It'll do your soul good.

Chapter 16

Let the Good Times Roll

A homestead offers pure enjoyment. Baskets full of eggs. Buckets full of sweet potatoes. Fresh milk foaming up in the pail. Animals dancing. In a day when mental health is on everybody's mind, what can be more healing than making beings happy each day and watching them respond gleefully to your care? I move chickens every day for functional reasons, of course. I want them on fresh pasture, away from yesterday's toilet—yes, function drives this.

But the beautiful part of this kind of function is how the chickens respond when I move them. They chase down grasshoppers, snip off clover leaves, grab grubs out of the ground. Part of my enjoyment is simply watching the chickens respond to my chores. Who gets to make this many beings happy every day? When I open the paddock gate for the cows, they literally dance. The calves gambol, cavorting around the pasture, kicking up their heels and swiveling side to side. Watching an 1,100 pound cow dance is sheer enjoyment.

How about the ripening grapes? In the spring, you prune them back and wonder how such a skeleton could ever produce grapes. You put on a fresh layer of mulch underneath and wait. You might sweat out that late frost, watching for the full moon. Grapes are pretty forgiving; even if the initial blossoms get frosted, they'll usually send out a second batch. The tiny, fragile blossoms open for pollination, and you watch the bees busily digging in from one to the next. Then the blossoms fall off, and tiny green nodules appear.

Over the next couple of months, those nodules get bigger, and the vines grow heavy leaves. What was a skeleton a few weeks ago becomes a tangle of foliage and bulbous grape clusters. Finally, they begin to color. We have concords at our house, and that first purple tint signifies not only ripening fruit, but the beginning of fall. Days are shorter, and nights have a temperature bite. Finally, the deep purple fruit attracts bees, and you know it's time to pick. You pick a few to taste and the deep, sweet resonance slides down your throat like nectar. Can anything be more beautiful? More enjoyable?

Picking day arrives, and you fill buckets and start squishing. In our house, we can the juice and then cut it in half with water for drinking. Few things are as beautiful as freshly canned jars of grape juice sitting on the pantry shelves. Lots of them. The grape vines, having given up their season's gift, quickly fade. The leaves turn brown and fall off after the first killing frost. The vines, naked and stark, await the pruner's hand and caretaker's mulch again in the spring to repeat the cycle. The vines at our house are now 75 years old. Beautiful. You can't buy that kind of entertainment.

What are other entertainments people buy? Gym membership? That's kind of a combination between entertainment and exercise. But on a homestead, you don't need a gym membership. Between digging post holes, working in the garden, shoveling compost, and bucking firewood, you get Crossfit without the subscription. Instead of spending time moving tractor tires from one side to the other, your exercise is productive.

Enjoying life, or what we might call happiness, has a mission quotient. I just had dinner with a friend who told me about an extremely wealthy acquaintance who committed suicide. He met with this businessman over lunch about a week before the suicide and still smarts over not picking up on the cues that indicated how serious his depression was. The thing my friend remembered the fellow saying was, "I have a lot of wealthy friends, and none of them is happy."

My friend, of extremely modest means, was especially taken aback when this wealthy fellow continued, "You're the only happy person I know." A week later, the extremely wealthy fellow took his life. That's tragic, but it speaks to deep mission. Enjoying each step in the mission is critical to having an overall positive disposition.

We all know money doesn't make anyone happy. I can assure you that when Teresa and I were starting out, living in the farmhouse attic, driving a $50 car, never eating out or going to movies, we were just as happy as we are today with a higher income. In fact, we've often mused about what we'd do differently if we ever really made money, and we've decided nothing would change except our charitable giving. Otherwise, we'd drive a used car, eat at home most of the time (we can't go anywhere and get better food anyway), and simply enjoy our place. A big part of my enjoyment today is watching our young people scurry around getting things done. Everything isn't on my shoulders anymore, which is a good thing.

Modern American living promotes the notion that everything worth enjoying happens somewhere besides home.

Too many folks view home as a pit stop between specific happiness activities occurring elsewhere. Even when we entertain ourselves for enjoyment in the home, someone else's cleverness and creativity are the centerpiece. We have home entertainment centers to import enjoyable content created elsewhere.

A home entertainment center is not a stage for us to come up with enjoyable content in the home. It's a place to pipe in enjoyment from outside. While I enjoy a good show as much as anyone, I would suggest that deep down soul-level happiness and enjoyment can never be satiated as much from outside as from inside. Laughing at a comedian's antics delivered on a screen certainly is enjoyable. But compare that to enjoying a beautiful garden bed of green beans. Or a root cellar full of freshly harvested potatoes and butternut squash.

One is enjoyment arriving from outside; the other is enjoyment that wells up from inside, from worthy work well done. Never has piping in enjoyment from outside been easier and cheaper, but I submit that never has accomplishing enjoyment from inside been more needed and less exercised.

After a day of Teresa's canning, which usually requires her to stay up late and finish off the last turns of jars, she usually sleeps in a bit the following morning. When I get up at my normal time and come down to the kitchen on my way out to do chores, I flip on the kitchen light and there, gleaming in radiant beauty, are 49 quarts of winter sustenance. Our canner takes 7 quarts at a time, so turns are in multiples of 7. Seeing those jars on the counter provides a level of enjoyment that no TV show can offer. One is simply recreation titillation while the other is the deep satisfaction that comes with meaningful work performed faithfully; the canning is mission-oriented.

Mission is what elevates activities to a place of enjoyment. Shallow missions don't reward with happiness at the level of deep and meaningful missions. When I look at a fresh batch of canned tomato juice, it's the reward for building the compost pile and spreading the compost on the garden bed. Starting the plants from saved seeds, tending them in the solarium on the end of the house. Then finally transplanting them in the garden and trellising them, mulching, weeding, and watering if necessary.

Picking, food milling, and canning yield an abundant and delicious provenance to sustain me during the winter. That batch of home-canned goodness represents not only compensation for work faithfully accomplished, but a cultural vote in the future. Teresa and I have defunded the bad guys and funded the good guys. We didn't put any synthetic chemicals or additives in that juice. It's just nature's goodness squeezed through a food mill. It contains all the essences of our soil. It's our worms. Our actinomycetes. Our protozoa. And that soil is now protectively preserved in the glass jars we've used for decades. No plastics. No landfill. Only a little lid to discard.

The jars aren't plastic. They aren't metal or tin cans. They aren't going to the landfill. This simple act not only feeds us, but it withdraws patronage from many things we'd like to see diminish. Like landfills. Like mining. Like energy for transportation. Like centralized food processing and distribution. All of this nuance flows through my mind when I flip on the kitchen light in the early morning's greyness and behold Teresa's late-night participatory faithfulness lined up in rows on the counter.

In that moment, my mind wanders to a cold winter night of ravaging snow. The radio offers cancellations like a ball game play-by-play. Sitting down to a warm, well-lit table, a large Revere Ware pot of steaming homemade tomato soup awaits a blessing and ladling into bowls. We'll sit and talk about the future, the past, and enjoy the moment. All these thoughts rush through my mind when I see those morning canning jars. That's deep joy, deep happiness, because it's mission-driven and ultimately meaningful.

In a day when the word defund seems to be bandied about as the ultimate test of ideological purity, how about we apply the word to things like trash, monocrops, and chemicals? Let's defund GMOs and factory chickens. Let's defund processed foods and unpronounceable ingredients. While we're funding our own enjoyable kitchens and homesteads, we defund outfits that cheat and cheapen. That's a great twofer.

How about relaxation entertainment? People go to great lengths to spend time relaxing in nature. On our farm, we have several picnic sites. We can have a picnic in the backyard, of course. Or we can have one by the creek. Or by a pond. Or on the mountain. You can spread picnic areas all over a homestead. "Where do you want to go for a picnic today, kids?" They have half a dozen options.

Most of the things people do to relax get them so stressed out on the way to relaxation that they don't relax until it's time to travel home. On a homestead, you don't have to wait for a once-a-year marathon relaxation time, called vacation. You can get out in nature in what I call relaxation commas. Too often, we think about relaxation in terms of length of time rather than frequency. I submit that more routine half-day commas (pauses) yield greater benefits than seldom long periods. Recuperation commas can be as short as a hot dog cookout on a fire or an evening spent reading aloud a classic novel. Short interruptions in work, rather than long multi-day heavily planned vacations, are much easier to schedule and therefore can be enjoyed more frequently. I'm convinced putting these non-work, non-hectic commas in our routine are critical to maintain life-work balance.

Ask any adult, "What are your most memorable times growing up?" They'll usually be times when they knew Mom and Dad weren't completely focused on work, chores, and projects. When everyone could simply enjoy each other in downtime. I'm amazed these days how many teenagers have never roasted hot dogs on an open fire. Or marshmallows to make smores. Talk about depriving children of what's important in life. Come on.

These kinds of home entertainment downtimes don't cost any money, except for the food, and you need that anyway. Why not do it outside on a campfire? And no dishes to wash. How entertaining is that? Eating out without the travel and hassle. If you get your hot dogs from an authentic farmer like Polyface (ha!)—or make them yourself—you can feel great about the food too. The marshmallows, well, we won't talk about that. I didn't say do this every night. An occasional spoonful of sugar won't kill you. The 80-20 rule is enough: 80 percent right and 20 percent compromised. That lets you enjoy your niece's birthday cake without guilt or being a party pooper.

If you're immune-compromised, of course, that level of going along to get along may not be acceptable. But for most of us, a bit of compromise for levity is diplomacy well placed. Party food is some of the world's worst, but every relaxation moment doesn't require party food. Grilled burgers are wonderful, especially over an open fire.

In the final analysis, relaxation is especially deep and

meaningful when our inner merit measuring stick indicates it's deserved. Few things are as vapid and undeserved as relaxation prior to accomplishment. At the completion of a big homestead project, dirty, tired, and hungry, relaxation reaches new heights of ecstasy. Our inner world and outer world align to offer restful moments that surpass shallow unmerited relaxation. When these downtimes mark accomplishment, like monuments to project completion, they create pleasure without remorse. That's a good thing.

How about recreation? Homesteads offer enough room to set up a volleyball court or even a paintball course. Think how much money people pay to play paintball. You could set something up and invite friends over. This could be a source of barter when you need to install some plumbing or electrical. Invite the children of plumbers, electricians, and welders over. Then when it's time to build your chicken mover dolly, you can cash in on your paintball course.

Do you need a thrill? Let me tell you, I don't need to go on rollercoaster rides or get scared at theme parks. When you cut a tree, and it falls into a nearby tree, and then you have to get under there and cut the supporting tree out, and run before the leaning tree falls on you—I've had enough adrenaline rushes for several lifetimes. You may not like me putting this in a chapter titled "let the good times roll," but everyone has different definitions of what qualifies as a thrill. For me, it's watching a tree fall right where I planned. Watching it fall where I didn't plan is a different kind of thrill. One is joy; one is terror. Kind of like a roller coaster.

If you really want a tree-cutting thrill, do what I did on one of our leased farms. Here at home, what we call "Polyface Central," we have power lines running through our front meadow. Our hundreds of acres of forest have no obstructions and I've never had to think about hitting power lines with a falling tree. At this new leased property, I cut a tree in customary nonchalance about power lines. Big mistake. Too late, I watched the top catch a power line and bring it to the ground. Wow, talk about fireworks. Sizzle, sparks and awe cut the power to a hundred homes. Oops. A frantic call to the power company and a couple thousand dollars later, all was well. Except my pride. Did I say something about good times?

Goodness, who needs to put their life on the line mountain climbing or cliff climbing, when you can get all the thrills you need getting a recalcitrant tree to the ground? Or catching an escaped rabbit? Or rescuing a baby calf from a flooded stream? A homestead provides enough thrills to put all the theme parks out of business. How about stepping on a rattlesnake? I've done that a couple of times. If I'd measured my jump, I expect I could have made the Olympic track and field team. Talk about heart palpations.

You get to do battle with rabid foxes. Few things are as scary as cornered predators. A big old boar raccoon is a nasty critter when he decides to fight instead of run. We have a perpetual control campaign on groundhogs; they dig holes big enough to break a wheel or upend a 4-wheeler. If you want an adrenaline rush, chase one down in the field and then miss it when you try to kick it. He'll turn on you gnashing his teeth and scare the bejeebers out of you. Who needs rollercoasters? Groundhogs are way less expensive and you can enjoy them right outside the back door on a homestead.

Probably nothing offers as much enjoyment as babies. About the only babies that aren't cute are mammals born hairless, like rabbits, rats, and mice. All other babies have a cuteness factor that's hard to beat. How many people get to welcome babies into the world routinely? To be sure, the ones that need no assistance are the most enjoyable. But if you need to assist in the delivery, and it ends well, that's pretty enjoyable too. If it ends poorly, that's no fun at the moment, but the tragic birth makes the others even more enjoyable because you know how tenuous the whole process is.

Watching a calf born, which is the biggest of our farm babies, is probably the most dramatic and most magical. The cow is big, and it's a big baby. Gestation on a cow is 9.5 months, and you can definitely tell when a cow is getting ready to freshen (that's what we call having a baby calf). She starts to swell in her rear end, and everything gets jiggly. Her udder also swells with milk; sometimes, she'll start to drip milk right before calving.

A normal delivery happens within a couple of hours, start to finish. The cow starts by finding a spot off to herself. As contractions start, she may exhibit a bit of agitation, getting up, lying down, getting up, lying down. This is all part of her feeling the calf and trying to get everything positioned for a smooth and trouble-free delivery. Once the water bag shows, things should progress rapidly.

I've watched many calves born and never tire of the sheer wonderment of the process. I've often thought about my feelings around watching a calf born and imagined how ramped up that must be for folks who deliver human babies. That would be everything I feel magnified by about tenfold. At any rate, within half an hour of the water sack appearing, the calf should be on the ground. Calves are supposed to come out with their head positioned between their front feet. A hoof shows first, then the nose and other hoof. By the time you see that, you're within minutes of everything being out.

With that final contraction, the calf shoots out and shakes its head to fling mucous out of its nose. If she's normal, the cow immediately turns around and starts licking the afterbirth off the newborn. Within minutes the calf is clean and beginning to wobble into a standing position. After a few failed tries, it stands, teetering, giving the mother a chance to lick off its belly and umbilical cord. Within half an hour of parturition, the calf starts looking for a teat. A good mother nudges the calf into position for nursing.

The pinnacle of enjoyment happens as soon as you see the calf nurse. At that point, the calf has a 99 percent chance of making it. Until then, lots of things can go wrong. But a standing, nursing calf is out of the woods. And then you see that tail wag with the first swallows of colostrum and a big smile breaks out on your face because you know all is well with this mom and baby. The fact that you picked that cow to breed, fed her, took care of her, brought her a bull, and now have the joy of a baby addition to your life makes the whole experience far more special than seeing someone else's calf.

If you've been there and done it, you know how special this is. Sometimes things don't go well. I've had those too. After waiting long enough for nothing to happen, you get the cow to the head gate and reach in to see what's going on. Usually the calf is in the right position, but it's either too big or the cow isn't pushing. She needs some help. Obstetric chains looped around the calf's front feet, and a pull, can bring the calf on out. If the calf is kicking and energetic, all is well.

Sometimes you wait too long or the calf is breach—birth is a time when a host of things have to go right. In any case, I've delivered dead calves. In such cases, I always let the cow go ahead and clean up her calf as part of her healing process—the nutrients from the placenta and afterbirth help her recover quickly. If I have an orphan I can put on the cow, that's paradise in this tragic situation. More often than not, though, the next day I take the calf to the compost pile. The cow gets turned back out with the herd to get fat and go into ground beef. We don't give second chances; that's the law of the jungle.

Again, the tragic ones simply make the normal easy births that much more special. After a few tragedies, your depth of enjoyment over the run-of-the-mill births deepens. The average person in America never viscerally participates in this level of life experience. When we see both the majesty and the precariousness of life this closely, we approach things more humbly, with less swagger. We realize our existence is special. The fact that our human delivery went successfully instructs our attitude when we've personally seen the highs and lows of life's beginnings.

With reinforced understanding of life's preciousness, realizing that not every embryo gets an opportunity to bring mission to the world, we see ourselves and others as anointed pilgrims passing through almost an ephemeral gift of life. Too many people are beaten down, concentrating on what's not perfect; let's realize that our very existence is a miracle. Then with gratitude, forgiveness, and savvy, we make of it what we can. A homestead encourages this kind of enjoyment and zest for life because homesteaders don't live separately from these profound intersections with meaning, purpose, and mission.

Homesteaders are immersed in the most mystical, magical, miraculous experiences of life. Whether it's a baby calf, baby tomato, or baby tree (we call these saplings—ha!), the sheer privilege of offering our hands and feet in a dance bigger than ourselves produces a special enjoyment factor. Every homestead has its tragedies, but if we stay with it long enough, we'll amass a long list of unspeakable enjoyments.

If I made a list of homestead enjoyment opportunities, it would fill many pages. Dams in the creek, sleigh riding down the hill, bonfires, hunting, discovering a clutch of ducks on the pond. These are enchanting things. Just a couple of months ago, some beavers moved into the river in front of our house. It's in a secluded bend, and we didn't even know they were there until we happened to see a two-foot high dam in place. All that happened while we were going about our business. That's what happens on a homestead. The most enjoyable things are often the ones you don't even plan. They just occur because you're in a place that's conducive to life, and life carries its own dynamic spontaneity . . . and hilarity.

I hope I've filled your head with all sorts of homestead home entertainment possibilities. From beauty to enjoyment, exercise to discovery, relaxation to recreation, a homestead fills all these human needs and desires for free, any time. If you never sell anything or make a dime from your homestead, having this level of life entertainment, experience, and stories beats all the planned, highly scheduled, expensive entertainment venues hands down. A homestead is where it's at, folks. Let's enjoy.

Chapter 17

Better Than Wall Street

Where do you put your money? If you're one of the few people with money left over at the end of the month, what do you do with it? That predicament affects all of us who have money left over. Realize that right now only half of all Americans can put their hands on $400. That's not much of a cushion.

Many of us still believe in savings, nest eggs, and accumulating some wealth, even if it's just enough to pay for elder care. As families break down and move apart, the old ways of elder care, using family, have largely gone by the wayside. What used to be considered family responsibility is now given over to professional caregivers who don't come cheap.

Debt-free guru Dave Ramsey has his multi-step program for wealth accumulation, but it still faces the conundrum of what to do with the leftovers. He says all will be well and to simply put it in the stock market. First, broad-based investment often requires some of my money going into dubious corporations. Second, these investments assume everything will rock along hunky dory for the long-term. Part of what's fueling the homestead tsunami is a growing concern that faith in Wall

Street may be misplaced. Maybe it'll continue for a hundred years on a gentle upswing, and maybe it won't.

National debt, funny money, unfunded government mandates, and many other financial issues make thinking people concerned about our nation's financial future. Increasingly, young people entering the job market don't see their trajectory as being as good as their parents'. Financially, what is truly dependable today?

For the record, I'd like to do away with money altogether, but that's not practical. The problem with money is it can be confiscated, manipulated, measured, or tracked by nefarious parties (government). Teresa and I have never found a definitive answer to the savings and investment portfolio problem, even though we've been blessed with numerous wealthy acquaintances and read plenty of books about finances.

My dad was an accountant and financial genius. He loved saving people money on their taxes and often got new clients by promising to save them more in taxes than his consulting and preparation cost would be. To my knowledge, he never failed to make good on that promise. After he and Mom paid off the farm mortgage, they purchased a three-apartment house in town as an investment. It required plenty of fix-up, but we pecked away at it during my high school years.

It was an old house near a college. Quiet street. Built into a steep hill, it had a basement apartment with ground-floor entrance, a second (main) floor apartment with ground entrance, and then a third-floor walk-up apartment. All three kitchens were on top of each other. One of them developed a leak and eventually softened the ground enough around that corner pillar on the downhill side that the house sagged a bit. Once we discovered the leak and got it fixed, we needed to jack up the corner and re-level the structure. My brother had already gone off to college, so I was the designated shoveler to dig under the crawl space and carve a trench for new footers.

We used massive house jacks and a railroad tie to lift everything a smidgen above level. Everything around the house was on a steep hill, which complicated logistics on the project. We couldn't get in there with a concrete truck because it wasn't flat enough to be safe for the truck to get close enough to pour. We ended up bringing a little electric mixer in to mix the concrete. We were able to back our truck in, with aggregate on it, but when we finished we couldn't get it out. We ended up calling a wrecker service to winch the truck out of the yard to the street. Did I say everything was on a steep hill? We wheelbarrowed—yes, uphill—the concrete to the footer. Once it hardened, we let off the jacks and everything settled into a perfect level position. I'm amazed I didn't suffer Post Traumatic Stress Disorder (PTSD) after that job. Wow; it still haunts me. I was 16 at the time.

We rented the upstairs apartment to two sweet-looking college girls who came with their gracious mothers to rent the apartment. All seemed well. But then the mothers went home, and the girls were there alone. Boy howdy. They had pets that peed and pooped on the hardwood floors. That's bad enough, but these girls never cleaned it up. Eventually, the lady in the middle (ground floor) apartment began complaining about odors, and Dad and I went in to check on things when we knew the girls would be gone. It was indescribable. The stench of dog and cat urine almost steamed up our eyes. We thought all those crunchies on the floor were dog food or pet food; no, they were hardened poop.

And the large porch, adjacent to the kitchen, had an entire

garbage truck full of kitchen scraps and trash. They had never taken any trash to the curb for three months. Dad and I toted all that stuff down to the curb—it literally filled a trash truck. Did I mention PTSD? As Dad always said, the property was a good investment and made money, but nothing could compensate for the emotional trauma of dealing with the renters. One basement apartment couple ended up being the local drug dealers. In an argument, the husband put his hand through the drywall. Nice guy.

Another renter had a boyfriend over, and in a drunken brawl, he tore down the screen door and the neighbors had to call the police, who called us at 2 a.m. We sold that house, licked our wounds, and looked at other options. That all happened while I was in high school. I know plenty of people have made money in real estate, but my experience with that house tempered my ambitions about being a landlord to folks I don't know.

Dad's two brothers had retired from excellent corporate jobs, had time and money to spare, and began investing it. Again, this was the early 1980s. Dad decided to try his hand at investing in stocks and bonds. This was back when you needed to work through a stockbroker to buy and sell stocks. I remember one day well. Dad and I were up in the woods chipping tree branches for barn bedding for the cows. We got in shortly after noon, and the phone answering machine was lit up with messages from his stockbroker.

Here was the message as well as I can remember it: "Bill, if I could have gotten hold of you at 9 a.m. you would have made $8,000, but since I haven't been able to get hold of you, you just lost $8,000." Dad listened to the message, looked at me and said, "I can't invest in something that fragile." He concluded that his brothers did well at this game because they were retired and hanging around the house all the time. This was before cell phones, computers, and do-it-yourself trading. We were outside working on the farm for extended periods of time; we couldn't be tethered to a phone in the house.

At that point, he decided to invest only in the farm, considering that the safest place. We'd never had the place surveyed; with 9 miles of perimeter, that was an expensive thing. A couple of neighbors were starting to do some logging next to us and nobody knew exactly where the boundaries were. That can be a big problem when someone starts cutting trees. He commissioned the survey, and by that time he was not well. I finished things off with the crew, and then Dad passed away in February, 1988. The stock debacle made Dad decide to only invest in things he knew about and could control.

Please appreciate that these kinds of experiences colored, and still color, my sensibilities about money and investment. I share them not to complain or dissuade anyone from investing in real estate or the stock market or any other investment you take a fancy to. The point is that investing in things outside our control is a bit of a crap shoot.

Now I'll tell you my stories in this arena. When each of us kids went to college, Dad opened a Whole Life Insurance account for us and paid into it until we got our first paychecks, then we took over. This was in the 1970s. During the inflationary 1980s, the investment and payout started to look unappealing, and I decided in 1987 to cash in my policy and invest the cash in a mutual fund. I had a friend who presented this strategy, and it made sense. Being young and healthy, I could get an inexpensive term life policy.

I still have emotional scars from the conversation with

the New York Life Insurance agent regarding my foolishness to cash out of that Whole Life plan. He turned hateful by the end of the phone call, making all sorts of dire predictions about where I was headed financially by making this stupid decision. I went ahead and cashed it in. Teresa and I took that money and immediately invested in a broad-based well-credentialed mutual fund. Do you remember what happened in 1987? In October of that year, within just a couple of months of making this gut-wrenching decision, we lost 70 percent of our investment.

Suddenly that New York Life conversation rang in my ears. Was he right after all? At the time, the plan was to take the premiums we'd been sending to the life insurance company and put them in a term life policy with the leftovers going into the mutual fund. We did the term insurance, but after being stung that badly by the mutual fund, we didn't trust it. Finally, ten years later, we crawled back to even, plus a little, with what we'd initially put in. The interest outpaced inflation by a fraction; we felt good that at least we hadn't lost anything. But it took ten years to get back to zero. It eventually started to gain in value, and we felt like we'd done the right thing after all.

You know what happened in the fall of 2008. That was the housing collapse and mega-bailouts to save Lehman Brothers, Fannie Mae and Freddie Mac, and all their cohorts. The collapse took our mutual fund back to our original investment. We were back where we started. By this time, we definitely didn't trust Wall Street and started looking around for alternatives.

I decided that Dad was right: invest in the farm. That's what we knew, where we had control, and if we were strategic about it, these investments would pay off better. Rather than putting money in retirement plans, investing in Wall Street, or buying gold and silver, we began a concerted effort at investing in the farm. We began targeting things that would make us more resilient.

Every time we had a few thousand dollars extra, we built a pond and added water pipe. Today we have millions of gallons stored in high terrain permaculture style ponds that feed a 10-mile network of gravity-pressure water lines all across our farm. If the power goes out, we have water. No pumps, no switches. And it's great pressure, at 70 psi. That's like a firehose. In the last several years, we've added enough storage to make a significant impact in irrigating during dry spells in the summer. That beats normal insurance policies all to pieces. While droughts may be as unpredictable as Wall Street, we know some sprinklers will fix a drought every time. When Wall Street goes dry, who knows how to fix it?

As I write this in January 2023, I have no idea where things are going economically in the world in the next year or two. Modern Monetary Theory (MMT), which essentially says the government can print money out of thin air as much as it wants with no consequences, is doomed to fail. I know that. When and to what extent, I don't have a clue. Maybe it will have already happened by the time this book gets printed.

I've never been one to make predictions, but I do think some rules exist. One is that whatever everyone thinks is the best investment probably isn't. We've just come through the FTX cryptocurrency debacle; I'm glad I've never invested in that. Another rule is that you don't want to be first, but you don't want to be last. Whatever the darling of the day is, you can be sure it won't pan out. The majority is usually wrong.

As we've seen pressure from the government to access pension funds and retirement accounts because they hold vast wealth, I don't put anything past government agents. Governments lie, cheat, and steal routinely—that's their business. I'm not even taking my Medicare entitlement because I don't want the government telling me what kind of doctor, hospital, or care I'm supposed to patronize. I'd rather walk away from the thousands of dollars the government stole from my paycheck than walk into a government-controlled healthcare system. As more people wake up to government-controlled healthcare problems, we're seeing a parallel health industry develop. From health sharing to cash-only doctors and herbal therapists, alternative practitioners enjoy robust patronage. My functional dentist can read my microbiome with an infrared light focused on my tongue. You won't get that from a dentist paid for by government programs.

With Central Bank Digital Currency (CBDC) at the forefront of monetary discussions and having already been largely implemented in China, who knows if you'll be able to access savings accounts if your social compatibility score isn't high enough? We live in tyrannical times. With all the advantages digital and e-boom money have brought, these technologies' ability to track, spy, and assimilate data is historically unprecedented. And downright scary.

Who will you trust with your money? Central banks? Wall Street? Mutual funds? Bonds? Treasury Bills? Gold? Futures commodities? I'm not a financial guru by any means, but I read widely, and I can assure you that each of these has a couple of downsides for every upside. Even gold can be outlawed. When you couple the disruptive forces in our country with the explicit goals of the World Economic Forum, it makes you want to go to a cave and hunker down.

I had a fascinating conversation recently with two about-to-retire U.S. Special Forces military husbands and their wives. They were convinced America was going to implode. If you study historical cycles at all, you know we're on the timeline for breakdown. All great civilizations have lasted about 200 years. If you look at the fall of the Roman Empire, America is following the same path. The iconic treatise written by Edward Gibbons in 1788 titled *The Decline and Fall of the Roman Empire* details why great civilizations wither and die:

1. The undermining of the dignity and sanctity of the home, which is the basis for human society.
2. Higher and higher taxes and the spending of public money for free bread and circuses for the populace.
3. The mad craze for pleasure—sports becoming more exciting every year, more brutal, more immoral.
4. The building of great armaments when the real enemy is within—the decay of individual responsibility.
5. The decay of religion—with faith fading into mere form, losing touch with life, losing power to guide people.

The cycle has been described this way by 18th century Scottish historian Alexander Fraser Tytler, known as the "Tytler Cycle:"

1. From bondage to spiritual faith.
2. From spiritual faith to great courage.
3. From courage to liberty.
4. From liberty to abundance.
5. From abundance to selfishness.
6. From selfishness to complacency.
7. From complacency to apathy.
8. From apathy to dependence.

9.   From dependence back again to bondage.

As you study these lists, where is America? Any observant person realizes that the institutions and thinking that brought our country to where it is either no longer exist or are in such a shambles as to not exist. When Rod Dreher wrote *The Benedictine Option*, he hearkened back to St. Benedictine's response to seeing Rome shortly after it was sacked. He realized the civilization was gone and would need to be rebuilt. Born into wealth, position, and power, his family lost it all during Rome's collapse. His family and friends had been grooming him to join the Roman senate. As a young man, lover of his country, innovator, and thinker, he returned to his home area and established a monastic order to preserve literature, agriculture, and religion. He envisioned a Roman re-birth.

Dreher and others today argue that the American civilization is already gone as surely as Rome's when Germanic tribes breached its grandeur. In vain do we attempt restoration at the federal level. This school of thought admonishes us to withdraw from the collapsed seat of power and start over, at the grassroots, with communities that preserve civilization. What we see emanating from Washington, D.C. is uncivilized. Trampled freedoms, thievery, corruption, and unprecedented partisanship threaten our ship of state dramatically enough to bail out. In doing so, Dreher argues we're then freed up to focus all our energy on developing civil communities, enclaves of civility, self-reliance, and artistic beauty.

We don't need taxpayer money to fund art exhibits portraying people urinating on Jesus. We don't need free drugs and sterile needle clinics to keep our addicts safe. What we need is the Ten Commandments reposted in school classrooms.

But alas, school classrooms have been taken over by an agenda of debauchery and anti-freedom that makes salvage impossible. Hence the rise in homeschooling and all kinds of alternative schools. As our nation's largest institutions, both public and private, bend to vengeful agendas, the homestead tsunami dares to invest in communities that espouse morality, freedom, and participatory self-reliance as a way to restore pockets of civility.

This investment is about creating places where life is safe, secure, and stable. Unfortunately, most of our nation's urban areas enjoy none of those privileges. As a nation, we've now tipped over to where more people receive their income from the government than those who pay the taxes. The farther we stray from meritocracy, the more resentment develops within the national psyche. To be robbed by a criminal is one thing; to be robbed by the government and told it's necessary to support laziness and debauchery is quite another matter. Those of us who work hard and see half our income absconded by the government to be wasted on outrageous expenditures are angry and frustrated. Rightfully so.

The ability and incentive of the government's hand to reach every bank account, credit card, and commercial transaction from payroll to investments are unprecedented. Powerful interests feel cornered by the economic mess they've created by overspending and debasing the money supply. "When cornered, a rat fights" applies not just to rodents but to bureaucrats, technocrats, and any powerful interests who want to maintain their hold on power and wealth. A fair question is how tyrannical and devious the government will become as it grasps for validity in a collapsing civilization. As Americans, we think surely our liberty DNA will constrain our government from doing things we've seen in other places.

But COVID dashed those hopes. Many of us now realize bureaucrats and many politicians feel no Constitutional constraints whatsoever. Any power, any action conceivable is fair game; the citizenry appears to desire control and tyranny. COVID created astounding penetration of government edicts into the most intimate crevices of our lives. Millions of Americans, for the first time, realized the government's capability and capacity to exercise its force on people. Unfortunately, most people went along. But as we think about our hard-earned money and how the Canadian government confiscated bank accounts from truckers who dared to question some of the public policy, it gives us all pause.

If you go onto websites that question societal orthodoxy, your search is tracked. If you make a phone call, use a credit card, attend a function—all of this is retrievable and trackable today. The bottom line is how to stash wealth and hang onto it in such unstable times. Where do you invest for the greatest return, the greatest assuredness that your wealth will withstand the shifting tides of a gasping culture?

Remember Y2K? Plenty of pundits predicted that at 12:01 a.m. on Jan. 1, 2000, the world would return to Neanderthal times. Airplanes would fall out of the sky; your car would quit running; the grid would shut down; water wouldn't come out of the faucet. None of that happened, but it does bear a lookback at what the folks who predicted the worst advised. And that brings me back to the two special Forces fellows and their wives mentioned above who called me asking about the best way to prepare for collapse.

These were sharp folks, and what they concluded was that they had two alternatives. One was to head for the hills and live off the land. People have certainly done that in the past, but you can't leave modern living and subsist in nature in a day. Learning how to trap, skin, build fire, make vessels to hold water, and all the other daily survival skills takes a lot of time. It also means your spouse and kids have to be on board. That's a big ask. If your strategy is to live in a cave in Idaho like a mountain man, you need to immerse in it now. You can't wait until things fall apart, after living all your life in suburbia, and then head to the hills and expect to survive successfully.

You have to develop skills and knowledge during years of preparation. Most people aren't willing to immerse that heavily in cave-dwelling lifestyles before needing to for real. We like our computers, takeout, and friends too much.

The second option was to get to a place where people know how to build things, grow things, and fix things. While the initial relocation can happen quickly, the relational intimacy required to make a community work takes time. That puts us back to the first option, the requirement to invest in the ark before the flood comes. As we chatted, the thing that became obvious to all of us—and I was learning along with them—was that the only real question is which scenario provides the most enjoyable preparation option and has the greatest chance of success?

Option two, the community option, seemed to offer the better chance of success. The two wives for sure liked that option better. They all thanked me for thinking through the conundrum and began planning where they would go. My point is that when we think about economic investment, we need to think way beyond simple dividends and cash returns. An all-encompassing return on our investment will be best served by placing our wealth in real stuff that has intrinsic value, like land, animals, plants, soil, and skill.

That sounds a lot like a homestead, don't you think? A corollary I like is the axiom "feed yourself first." You can't help someone else unless you're healthy and have some wiggle room in your own life. Every flight attendant goes through this in their safety briefing: "put your own oxygen mask on before helping anyone else." Why? If you're drowning, you can't keep someone else from drowning.

Investing in your own food and your own ability to eat surely makes sense as much as trying to jump on the latest venture capitalist train. When you weigh options like Wall Street versus your own food, water, and warmth (like firewood), it's not a question of evil versus righteous. It's a question of importance and value. Which one is more basic, more valuable to preserve life?

For the record, I don't think it's wrong to invest in all the things I've mentioned as being tenuous or even having negative aspects. But reason suggests we should invest first in the most fundamental guarantees of life's flourishing. Once that's done, do whatever strikes your fancy. Cryptocurrency, mutual funds, gold—a diversified investment portfolio always makes sense. Teresa and I have invested in local business. But that is after life's fundamentals are battened down and relatively secure.

Wendel Berry writes that what is wrong with us creates more Gross Domestic Product (GDP) than what's right with us. He points out that if you're happily married, living on one income, growing your own food, and creating your own entertainment, you can live on little money and you don't add much money to the GDP. But if you get divorced, now you occupy two houses instead of one, need to work two jobs instead of one being enough, drive two cars, eat meals out because you don't have time to cook, and buy all your food because you don't grow it any more—all creating extra economic activity and adding to GDP.

Throughout this book we've touched on numerous places that have economic impacts in life. A homestead provides a wonderful platform to develop a positive self image. Kids with a good self image don't need to visit expensive psychologists or take anti-depressant drugs. These are real economic costs for a family. I meet many folks in their 30s who suffer immune dysfunction and debilitating diseases due to poor food, lack of exercise, and emotional trauma from growing up in a stifling urban habitat. After moving onto a homestead, they heal, and their kids exude joy and confidence.

In October 2022, sports gambling generated more than $500 million in Virginia alone. Think about that figure for a moment. That's just one month in one state. Read the real estate mansion section in *The Wall Street Journal* in each Friday edition. Folks, the world is not starved for money. The world has plenty of money to solve every problem we can imagine. We don't suffer for lack of money; we suffer for lack of spending strategically on the most important things in life.

I can hear the naysayer responding, "Well, everyone can't buy a homestead." To which I counter, "What about you? Everyone isn't, so you don't have to worry about that. Everyone didn't go to the ark; it was just a handful. What are you going to do? That's the only question." In other words, if nobody else wants to don a lifejacket in a sinking ship, will you refuse to put one on too? I know well enough that everyone won't invest in a homestead. But that doesn't mean it's not a good investment. I have no delusions of grandeur that this book will suddenly spawn a bigger homestead tsunami. Come to think of it, I think you should buy a box of these and give them to all your friends.

Maybe we can in fact elevate the tsunami. Ha!

Dave Ramsey's financial radio show receives countless calls from people lamenting that they can't make it. They're in debt over their eyeballs and sinking. His first question is always, "What do you earn?" More often than not, it's way more than $100,000 a year. For most of our lives, Teresa and I have lived below the official government poverty line. But we've always been happy, eaten well, and stayed warm in the winter. Most years we even have cash left over. When you do a lot of your basic living cashless, you don't need to earn much. Then you pay fewer taxes. Then you defund the government. I like that ancillary benefit. I can't think of very many entities more worthy of being defunded than the government.

Could the government take away your homestead? Perhaps. But compared to all the other things you could invest in, our domiciles are probably as secure and stable as anything you can imagine. Our house and property will be the last bastion of freedom and private ownership when all else fails. I don't even want to think about what happens when our homes get taken. Let's build enough and love our properties enough to bring enough people to our side to have enough societal credibility and attractiveness to contribute help and hope when others are helpless and hopeless.

Rod Dreher's Benedictine option makes a lot of sense when we find ourselves increasingly frustrated and depressed about the state of the federal government and big institutions. As the saying goes, why adjust the deck chairs on the Titanic? Find a lifeboat and jump in. That lifeboat might be called a homestead.

A homestead as a pure economic investment is one of the best reasons to make the jump. Welcome.

Chapter 18

Finding Friends

Running deep within the DNA of Americans is something called rugged individualism. As a culture, we pride ourselves in not asking permission, doing our thing, not feeling any need to be endorsed or affirmed by anyone else. "You do you, and I'll do me," we say, usually when someone has questioned our activities. That response creates a lot of latitude for stupidity.

This societal persona did not develop only because our nation had a love affair with liberty. Individualism on steroids developed because we're a nation of immigrants from every part of the world. Some people came of their own accord; and others (enslaved people) did not. But the people who forged the persona of our culture came from many other places. Yes, they overwhelmed and certainly mistreated the Native Americans, who had fought and warred among themselves for centuries over territory and power.

Conquest, mistreatment, and subjugation did not invent themselves in the American experiment. These horrific activities constitute most of human history, in all parts of the world, with all sorts of cultures. In that respect, people are people and nations are nations. Humans fight over resources; that's what we do. Nations are simply extensions of individuals, making national fights bigger than individual fights. America isn't different; it's made up of humans.

Where America is different is in its immigrant founding. Yes, people have migrated since the beginning of time, but never from places as divergent as here. Occasionally we see regions settled by old-country people groups. Places like Chinatown. Or Norwegians in Minnesota—I guess they like the cold. But those enclaves are the exception, not the rule. While the overriding immigrant population was from the countries of the United Kingdom, waves of people came from other places.

I've traveled in Europe a fair amount and am struck by their cohesiveness. These old cultures have characteristics that deeply define them. I stayed in a hotel in Amsterdam that was 500 years old. When people live together in a place for that long, through wars, political turmoil, famines, and floods, they become much less individualistic. They value cohesiveness. Their communal collective agreement, culturally, is unlike anything we have in America.

Even in America's tight communities, like churches or civic groups, members nurse a bit of a rebellious nature. We subconsciously yearn to be the maverick, to go our own way, and do our own thing. As time beats people down, the alternative view becomes less attractive. The way things work, how we do things, accepted practice, becomes axiomatic over time. Extremely old and less diversified people groups tend to adopt a don't-rock-the-boat attitude.

That's the way I explain the European aversion to genetic engineering and many food additives ubiquitous in America's food supply. In these extremely old cultures, even their governments seem to have a more benevolent and caretaking mentality. As if they're safeguarding precious legacy that would be jeopardized if too much individualism reared its ugly head. This cohesion is not about centralized power as much as it is about homogeneous custom.

I have a great friend in Norway whose family owns land dating back 500 years. Although never royalty, they were nobility and cavorted in high circles. The family still owns tens of thousands of acres, and I asked him how they held onto that land for centuries. In the U.S., farmland is broken up routinely through equitable inheritance structure. The American egalitarian mystique intervenes in inheritance, giving each child equal portions. Parents equate assets with love; you can't give one child more assets than another lest you be accused of having favorites.

As a result, a scourge on America's agricultural face is equal inheritance among farm kids. Invariably, one loves the farm, but the others don't. An "equal but undivided interest" means the child who wants money instead of land can force the others to sell in order to turn the land into cash and pay off the one sibling. As a result, American farmland routinely gets broken into smaller and smaller units, making commercial full-time farming more difficult. Preserving big chunks of land is a sign that a nation's agriculture policy is sympathetic to viable farming.

To hold onto such a massive parcel for several centuries seemed almost other-worldly to me and prompted me to query him on how it could be done. Without hesitating, he responded, "primogenitor." I blinked. "Say what?"

He laughed good naturedly and explained the European custom, dating back to Biblical times, that required giving the firstborn either all or most of the family land in order to keep it from being broken up. This is why European artisans, including mercenary soldiers (remember the Hessians captured in Trenton, New Jersey during the Revolutionary War?), were second and third-born sons. The first-borns got the land.

Noting that my friend was the second-born son, I couldn't resist asking, "How does that make you feel, as the second-born?" Without a hint of resentment, he answered, "No problem. My whole role in life is to make sure my older brother doesn't lose it." It was such a charitable yet matter-of-fact answer I was dumbfounded. Upon later contemplation, I realized the social equity, the familial power, behind both the statement and the practice. This fellow grew up in a culture where legacy trumped equity. Where the family's interests outweighed individual aspirations. Done at scale and ubiquitous in the culture, such thinking creates social cohesion Americans can only imagine. You only find thinking and contentment like this in extremely old, stable, homogeneous cultures.

King Charles III routinely says that a culture is known by its architecture, religion, and food. Most cultures have distinctions in all three of these areas. America, not so much. We're truly unique in this regard, and it doesn't take much traveling around the world to notice. I confess to this rabid individualism, subconsciously applauding myself for being a nonconformist.

But let's look at the other side of this coin. Military personnel wear uniforms. Why? They don't applaud soldiers who wear different clothes to inspection. The whole idea is to look like everyone else. Why? Because they're one unit, one group, one purpose. One of the quickest ways to get students focused is to require them to wear uniforms in school.

Suddenly they're not distracted by wild and whacky self-expressed clothes. Uniforms hold students' shock value desire in check, forcing them to express themselves in character and attitude rather than threads. Every school I know that requires uniforms excels academically and socially beyond those that don't. They have fewer mental problems, fewer disciplinary issues, and much greater school spirit. Fashion shouldn't be the foundation of notoriety.

Perhaps the most distinctive people in America are the Amish and Mennonites, especially the horse and buggy groups. I've had the distinct pleasure of spending quite a bit of time with the Amish, and while they don't have everything figured out, I covet many of their community-oriented protocols. They take care of their elderly and are exempt from collecting social security (FICA) taxes from employees. They don't buy insurance. If a house burns down, everyone assembles to clean up the mess and build a new one, often completing the whole re-build in two or three weeks. Amish barn raising events are legendary—they look like ants on the structure, often completing the entire building in a day.

The burned out family is back in a new house before the rest of us could even get an appraisal of damage from our insurance company. They have their own schools, sheltering their children from society's idiotic values. When they travel, they are welcome in anyone's house. Perhaps no story illustrates this better than what happened to us on one of our field days many years ago. For 20 years, our farm hosted a national field day. It finally grew too big to handle and we discontinued it.

I'm the extrovert; Teresa is the introvert. With 1,500 people traipsing over the farm on one of these galas, she'd had about enough and headed to the house for just a few minutes of peace and quiet. And to visit the bathroom. People energize me but they wear on her; that's classic personality type stuff. Imagine her surprise and chagrin to come into our house for a respite, only to find half a dozen Amish moms sitting on our bed nursing their babies. They found a quiet corner too and thought nothing of using our bed as a breastfeeding staging area.

The next time we did the field day, she posted a sign on the front and back doors: "Private Residence. Keep Out." I confess that after spending a delightful day with the Amish, I chafe at the power of the bishop and the minute detail in their rules and regulations. As a group, they relish diminishing the individual and increasing the community's homogeneity. I know they struggle holding the line against technology. Their interface with the rest of society is often ragged and a bit blurry. The bishops have a hard time keeping up with technological advances and reining in their folks the right way at the right time.

If the rugged individualism American style represents the far end of self-interest, perhaps the community living Amish style represents the far end of communal interest. The balance is probably somewhere between the two. Actually, I spent time at a commune once that was over the top—at least for me. The common dining hall had a huge credenza on one wall with designated clothes cubbies: "Women's Panties, L," "Men's Briefs, M." I kid you not. Talk about share, and share alike. I don't think I'll join that group any time soon.

Since most Americans are not Amish, I'm concentrating on the weaknesses of over-individualization. Our nation's worship of the individual, almost to the exclusion of the neighbor, sends an imbalanced signal to our kids. The result of such me-centric thinking is an arrogant attitude toward others, a disrespect toward diversity, and an incorrect perception of our own competence.

Numerous surveys and studies in recent years show a widening gap between how American kids perceive their achievement versus what they actually achieve. Among developed countries (I don't mean anything condescending by that remark—it's just a way to differentiate; I could say rich, but that could be more offensive), ours ranks last on most academic metrics. But when surveyed, our kids think they're at the top of the heap.

Of course, the foundation for community is the family. Without strong families, we can't have functional communities. When one parent is missing, everything becomes more difficult. When everybody has their own room, monitoring is harder. And now with screen devices in everyone's pocket, divided interests and lives are easy even though everyone sleeps at night under the same roof. Kids get into drugs and all sorts of things without the parents even being aware. Kids turn into terrorists, and nobody knows. Sleeping several to a room, without private corners or quarters, encourages accountability. That's healthy for cohesion and family functionality.

Yes, Amish kids sometimes go off the deep end too, but generally everybody knows who's going off the deep end. Without a million distractions to suck individuals into silos, group conversation and interaction make becoming a terror harder to hide. Nothing is perfect on this side of eternity and nothing can be. We're looking for balance somewhere between the extreme of rugged individualism and the extreme of complete conformity.

A society bereft of individuals who dare to question

orthodoxy would never enjoy innovation. A society bereft of communal responsibility is like a bunch of loose cannons. Crisis tends to bring communities together. In that vein, I have a friend who, several years ago, started what he calls a Mutual Assistance Group (MAG). It's simply an informal group that meets quarterly to work on community resilience. They encourage the attitude of "I've got your back."

For example, during COVID, if someone lost their job due to refusing to get jabbed, the group passed a hat and picked up living expenses for the family until another job could be procured. As you can imagine, they discuss self defense. They talk about what they would do if the grid went down. If water suddenly became poisoned or unavailable. If they couldn't get gas or diesel. I think every community in the country should form a MAG to come together as a local group to prepare and help each other through hard times. The group is completely non-sectarian, non-partisan, and non-censoring. It's like Amish without rules.

Throughout history, people have spent more of their lives in hard times than easy times. Here in America, we've had easy times for a long time. Chances are those easy times will end; that's the way to bet if you study history. What happens to society then? Thinking people put some effort into contemplating these things. As Indiana University's legendary basketball coach Bobby Knight used to say, "Everybody wants to win. But are you willing to prepare to win?"

A friend in New Mexico is involved with an outfit called Lighthouse Pioneers that has created an exam template to measure functionality in twelve areas. They call them pillars:

Faith

Water

Food

Health

Shelter

Education

Economy

Relationships

Environment

Energy

Community

Transportation

We're challenged to rate each of these pillars as being in survival mode, resilience mode, or freedom mode. This matrix is wonderful for determining where we are as individuals, families, neighborhoods, church groups, or a nation. Obviously, the ideal report card would put each of these pillars in the freedom category, not just survival category. The goal is to figure out action steps and strategies to move all of them to a freedom designation.

The fact that, in my little bubble of life, I'm seeing these kinds of efforts popping up around the country indicates a hunger and yearning to create communal strength as preparation for national weakness. To be sure, that weakness may follow a time of unprecedented tyranny—that's usually the case—but having all these pillars in the freedom column would be beneficial regardless of the societal scenario.

With all this in mind, the homestead is the best incubator to balance individualism and community. I've visited many, many homesteads, and each one expresses the talents and interests of its designer. Some are extremely neat, even manicured. Others are a hodgepodge of projects, half-finished, and pieces of metal and lumber lying around. Some homesteaders put great emphasis on flowers, and others put all their energy into animals. No homestead is identical to any other, just like people. They literally take on personalities.

But homesteaders, as a group, are the most community-minded people I know. Willing to share and often needing help, homesteaders living near each other form a perfect picture of mutual interdependence. One of the biggest mistakes beginning homesteaders make is holding an unrealistic objective of self-reliance. They read the books and watch the videos, coming to the new paradise property with dreamy eyes and starry visions of doing everything.

Invariably, within a year or two, they're bogged down in a quagmire of unfinished projects, putting out fires and trying to survive each day from crisis to crisis. If you're in that frustration stage, or even at burn out, stop doing everything. You don't have to be cultishly independent. Nobody can be. What you need to cultivate is mutual interdependence. As you turn self-reliance into shared-reliance, the burden of having to do it all yourself will ease off your back, and you'll feel a renewed energy to tackle a few things well.

Education, which is a close cousin to experience, is both financially and emotionally expensive because it entails a lot of failing. Laser-focusing on your educational priorities protects you from bumbling along in everything. Select your priorities carefully, based on need and interest; if the need and interest are compelling, you'll probably figure out the skill of it. Once you achieve one success, you'll be more enthusiastic about tackling the next one. When everything is skating on the edge of catastrophe, you feel like you're in the midst of a fire without any marked exits. That's when you suffer burnout.

Step away, re-evaluate what is most important, and pare your list down to a couple of things. Growing five vegetables well is better than twenty barely surviving. Determine to get one bumper crop from the one thing you find easiest to grow. With success under your belt, you'll have an emotional snowball propelling you through other projects. If doing too much at the beginning is the biggest mistake for new homesteaders, the second is similar: not asking for help.

Swallow your pride, and ask for advice. Ask lots of people for advice. Don't assume the first person that gives you counsel is the final say. You'll get different ideas from different folks. You'll never go wrong having more ideas on the table. Enjoy the fact that you have a neighbor for a mechanic. Enjoy the quilter next door. Intertwined relationships create intimate friendships, which then create commitments to care and love. You can't help someone you don't know, and you can't know someone without doing things together.

The foundation for this mutual interdependence is sharing. Expertise and infrastructure are the two most common candidates to share. Again, I can't help but point out what happens in an Amish barn raising. All of us have seen pictures of 50-100 Amish crawling all over a new barn structure, taking it from start to finish in a day or two. It's truly remarkable and forms perhaps the linchpin in their social structure. From the outside looking in, the Amish barn raising is both a product of their community and a catalyst to their community. Which comes first is not clear. We do because of what we are; but we also are because of what we do.

Homesteaders tap into that same benefit. For many years, our closest neighbor in proximity was a master mechanic. Every time we got stumped in the shop, we'd go down to his house with the part or the piece of equipment and ask him what to do. If it was too big to transport, he'd come down to our shop. Once he retired, he came down several days a week just to hang out with all our young people and enjoy the activity. He's been deceased for many years, but hardly a day goes by that we don't remember some shared expertise experience with him.

The first time we butchered a beef in the pasture, Dad called another of our old-timer neighbors over to help. He loved to butcher and showed us how to block up the shoulders to field dress such a big animal without hoists. We'd never done it before but after that, we confidently did more on our own. When he got too old to get around, his family would bring him over when we butchered chickens. By that time, he was blind from diabetes and could hardly hobble around, but he could hear and talk. I'm confident having him close to our chicken butchering provided him with some of his most end-of-life joy. Meanwhile, we'd ask him about the old days, and he rose to the storytelling, regaling us with nostalgic memories and the history of our area.

We learned our community's history, the history of our own land, and lots of things people did long before we arrived on the scene. For much of his life, he would get up at 4 a.m. each day, and hand milk 10 cows for Grade B milk, then go to work in a factory, and then milk his cows again when he got home. His massive hands and fingers indicated a life of strenuous exercise. I had the privilege of helping carry his casket to his final resting place in the Lutheran Church cemetery.

Another of my old-timer neighbors showed me chainsaw cutting techniques. He had an old Frick sawmill when we moved here in 1961 and had spent half a lifetime working in the woods. He could roll a log with a cant hook and make it look like he had a front end loader attached to his arms. I could go on and on about these wonderful neighbors. Why did I learn from them? Why do I tear up remembering them? Because we shared. We shared expertise, equipment, and tools. But most of all, we shared meaningful work and noble projects.

Isn't it interesting that the sharing economy is presented in business circles today as something brand new, that's just been enabled by the internet? We can now share rides through Uber and hospitality rooms through Airbnb. All because of the internet. I've got news for you. The sharing economy predated the internet by a long shot. And that sharing created deep, deep friendships centered around place and purpose.

As political reporter Thomas Friedman pointed out in one of his books, everyone in the world is only one-sixth of a second away from everyone else through the miracle of global communication systems. That's how long it takes an electronic signal to go from one side of the globe to the other. But that one-sixth of a second is too far removed for a kiss or a touch. Homesteaders, as a group, bring back to our culture the social structure of yesteryear's visceral sharing and the intimate touches it engenders.

Compared to the shallow, fleeting, conditional relationships fostered on social media, sweating together, pondering a broken fence post together—these are the things that bind people into unconditional relationships. These create lifetime commitments. You can't afford to be easily offended when your life depends on your neighbor. You can't afford to be prejudiced when you need your neighbor to hold the light while you help deliver a calf at midnight.

As our society devolves into partisanship, censorship, and hatred, homesteaders uphold functional social structure partly because they like it but mostly because they need it. Everybody needs to feel needed. We all know that. And yet our urban gravitation toward individualism, interacting with people only on screens to avoid eye contact, reduces needfulness in our daily lives. Where we feel most needed is on the job, but even there we know we can be replaced in a heartbeat. So how needed are we in that replaceable situation?

On a homestead, if I don't feed the chickens, nobody else will. If I don't milk the cow, nobody else will. And if some crisis occurs and I can't handle chores, a friend or neighbor will step in and help because at some point that person will need me. No pay. No expectations except human loyalty. No responsibility except caring for what needs to be done. That's the aura that surrounds homesteaders and why this homestead tsunami is the most stable repository of social structure. If you're yearning for real, unconditional, lasting friends, a homestead may be the best fulfillment.

Chapter 19

It'll Be Okay

What do homesteaders show off? They don't take pictures of their clothes (fashion). They don't take pictures of their equipment. They don't take pictures of their houses. The lion's share of their pictorial pride centers on something they produced.

You see a child nestled in the midst of baskets and buckets of freshly picked vegetables. Perhaps a wide grin smeared with mulberry juice. A basket of eggs. Animals of all kinds. A woodpile mountain. Homestead talk focuses on abundance. It's all about provision and blessing.

Sure, talk to any homesteader long enough and you'll hear stories—and maybe see pictures—of floods, withered plants, weeds, bugs, and dead chicks. But those are generally a parenthesis on the way to the main story, which is all about blessing.

If a homesteader's synopsis is negative and depressed, it reflects more on strategy, personality, and expectations than on inherent weaknesses with the property. If you're going to do the homestead, it means you can't do other things. Like run the kids to little league every day. Like call out for pizza delivery.

Like extensive travel—especially if you have animals.

It's kind of like involving yourself in a local fellowship group (yes, I know we normally call this church, but I try hard to emphasize people and not the physical structure, so please allow me to use the phrase local fellowship group). If you're there on Sunday morning, you can't be at the beach. Managing an internal or outreach ministry might mean you have to work your schedule around responsibilities to give you time to oversee that responsibility.

Life is full of choices like this. We all have 24 hours a day; nobody has more. None of us can do everything we might enjoy doing. We have to prioritize. If we're going to serve God, for example, we can't equally serve competing interests. When we choose to adopt a homestead, its infancy will be full of poopy diapers and sleepless nights. Everything new has an infancy period that takes a lot of babysitting.

Over time, your homestead will grow into maturity and things that used to be frustrating will be second thoughts. You'll find previous struggles turn into effortless enjoyment. In a new marriage, you have a lot to sort out; fortunately, that coincides with the honeymoon phase. The early romance helps you plow through the getting acquainted cycle until you settle in to a more comfortable knowing each other phase. You learn what irritates and what pleases; all that takes time.

Much later in a marriage, it moves into that mature, deep love state. With all those fits, fights, and fun times you have treasures of experience that bring on soul-level appreciation and gratitude. "You've loved me this long. You've been my loyal partner through thick and thin." It may not be sparkly fireworks like it was on day one, but you feel deep contentment holding hands on the sofa or sitting close on the porch swing. Despite what our modern culture may say to the contrary, connecting emotionally is far more satisfying than connecting sexually. That deep connection takes a lot of seasons to cultivate.

Homesteaders go through the same maturity process because homesteads are not dead; they're living things. I know the house boards don't bruise when bumped, and the shovel doesn't blow kisses. But a homestead as organism exudes life, and as caretakers we're there to tease out its best performance. In the midst of tragedy, we imagine success. In the midst of loss, we imagine abundance. That's why we came here.

Like any marriage, fellowship group, or philanthropic organization we join, homesteads require dedication. They're jealous for our time and attention, kind of like God, who says He's jealous. God doesn't like divided loyalties. You can't serve the homestead while spending all your time dreaming of travel destinations. No, you've married a homestead. That's your destination; it's your anchor. You have responsibilities now.

And if you get your priorities straight, you'll be able to take those pictures of abundance homesteaders love to take. The cow will calve a healthy, robust next generation. The butternut squash will cover the ground, and late in the season as the vines die back, reveal wheelbarrow loads of succulent dinners. The children will love the dishes cooked with fresh herbs from the kitchen garden. Friends invited over will marvel at how good everything tastes.

What I'm suggesting here is that just like the prospects of marriage require faith that it will work out over a lifetime, and just like joining a fellowship group requires faith that this is a good investment of time and energy, and just like our faith in God requires us to live a life pleasing to His principles, so homesteading requires faith: in soil, plants, and animals. It

requires faith that in the lowest times, things will eventually
work out. Faith that day will break upon night, sunshine upon
flood, abundance upon scarcity.

This is why homestead health doesn't center on drugs and
vaccines. Homesteaders, by and large, have faith that if they
honor and respect the pigness of pigs, providing a habitat that
allows them to fully express their pigness, they will be healthy
and happy. The conventional industrial farm doesn't ask how to
make pigs happy. It only asks how to grow them fatter, faster,
bigger, and cheaper. That question forces hog farmers into all
sorts of convoluted thinking about medications and provisions.

By faith, mature homesteaders know that well-cared-for
animals don't always reward with health. Sometimes, despite
all our efforts, an animal dies. A pepper plant wilts away. A
cucumber vine shrivels with powdery mildew. It happens. But
by faith, we keep trying. By faith, we look at the healthy ones
and realize the sickness is aberrant. Just like, by faith, we dust
ourselves off after a marital argument and carry on for the
future. Setbacks are merely learning sessions on the way to
mastery.

We don't view nature as a reluctant, petulant partner to be
wrestled and pummeled into submission. We view nature as a
benevolent lover that wants to be caressed in the right places.
This is why permaculture teaches a one-year get-acquainted
period with your land before doing anything. That way, you'll
know where the snow piles up; where the winds blow; where
the brambles proliferate; where grape vines trellis into the
trees. Each homestead is a unique place, just like a person. It
has a climate, character, and context that define the best way to
partner toward success. You don't want to grow an orchard in a
frost pocket. It's not about what you want; it's about how God

made that spot. Our stewardship is learning about the land's bent and then massaging it toward its natural proclivity.

This homestead faith requires placing our trust in functions humans didn't design or build. We don't know why earthworms turn one way instead of another. We don't know half of what goes on in the soil. We don't know why one chicken gets up earlier than another. We don't know why chickens like to rise early, cows a bit later, and pigs sleep in. We don't know why one apple is bigger and its twin smaller.

But we have faith in an infinitely complex and symbiotic system. That faith makes us confident that supplying compost around the base of the apple tree will make healthy, robust fruit on average. That mulching the garden preserves moisture, encourages earthworms, and brings healthier yields to our harvest buckets. That moving the cows around the pasture like the bison moved on the prairie duplicates nature's template in microcosm. And it works.

I believe God made the physical world as an object lesson of spiritual truth. We humans have a hard time understanding things like provision, forgiveness, mercy, even love. We struggle to hang visceral understanding on a mystical set of attributes. And so God gave us cows. The cows love to be cows. They're herbivores, not carnivores. They're curious. They're big. If they're not where they're supposed to be, we don't get angry at the cows. We look to ourselves and ask why we left the gate open, didn't fix the fence, or let the electric fence energizer run out of power.

The cows are there to teach us about ourselves. Are we lazy? Negligent? Too busy to care for what we've chosen as our responsibilities? Unobservant? Did we scream at the cows in anger when really they were just being cows? God made things to be and do in certain ways. A cow that wasn't curious about the other side of the fence wouldn't be a cow interested in finding the most palatable blades of grass. At the same time, the cow that jumps a good fence and is an incorrigible rogue needs to go in the freezer in delectable little packages.

Patience and longsuffering do have an end. And some cows can help us understand that too. The Bible's Old Testament has a lot to say about rogue animals. It even says that a pushy cow, known to be pushy, that injures someone is not only responsible but the owner of that cow is responsible because he let it continue to live knowing it was pushy. A homestead offers an object lesson for faithfulness in training, stewardship, and patience. Homesteads provide the ultimate classroom and tests for our faith journey.

The way I view my animals and plants tells a lot about the way I view people. If I see pigs and tomatoes as just piles of protoplasmic structure to be manipulated however cleverly my hubris desires, chances are I'll see people the same way. A culture that doesn't think happy pigs are important probably doesn't think happy people are important. Our attitude toward the least of these hangs measurable standards on our attitude toward the greatest of these (people). To be sure, that doesn't mean we worship animals or worship creation.

It does mean we caretake what is not ours to bring glory to God, who owns it. That caretaking is part of our worship of the Creator, not to be confused with worshiping the creation. We love to dissect our lives into the sacred and secular, but homestead stewardship helps us understand practical sacredness in seemingly mundane activities.

Ultimately, homesteaders learn to dance with a partner partly of their making and partly other-worldly. Every vegetable grows differently. Ten plants started from the same seed packet will not be identical. They can be equally productive but will not have the same form or branch pattern. This is a beautiful picture of diversified abundance out of common stock. Just because a plant looks different than another in the same garden bed doesn't mean it's unproductive. A person who looks different from me can produce all sorts of good ideas and contributions.

Interacting with a menagerie of plants and animals exhibiting this kind of diversity touches our subconscious to appreciate differences among family members and people in general. Homesteaders know that productive cows come in all shapes and sizes. Sometimes the cow with the crazy horn is the best milker. Or the cow with the crazy mosaic on its side is the most gentle. Or the cow that's hardest to keep fleshy is the most dependable for breeding, calving, and having a healthy udder. Sometimes the prettiest cow is the one with the most mastitis trouble.

Isn't that just the way life is? God's humor is all over a homestead. Often the cutest calf is the one that grows up to be the barren one. The less pretty often becomes the faithful old cow that is the most productive. Outward beauty doesn't always show the final thing. By faith, then, we don't pick everything based on outward form; we let function determine the favorites. That sounds a little bit like Jesus saying we know Godly people by their fruit. I've always said an egg a day keeps the hatchet away. Bearing fruit is perhaps the ultimate test of authentic faith.

Few things depress a farmer like drought. Flood is certainly catastrophic, but it's normally short-lived. Droughts, on the other hand, can last for a long time. Each day you look at the sky, hoping to see black clouds forming in the distance. But every day the sun bears down, sucking more moisture out of the parched earth. The soil cracks. Grass withers. Day after day after day. When I was younger and starting out farming, these drought times would send me into a deep funk. I couldn't do anything about it. Oh yes, I'd pray, but in the end I knew rain was completely out of my control.

God even says it rains on the just and unjust, so apparently He doesn't even manipulate it much. This is where the Deists got their idea of the earth being a clock. I don't believe the earth is alive like the Gaia concept, but I also don't believe it's a simple hunk of rock. It has dynamic forces playing on it, not the least of which is the moon. Magnetic fields, heating and cooling through water vapor condensation, or lack thereof. Lots of things are going on. Sun flares. Goodness, we can scarcely predict the weather a week out, let alone what will happen next year.

Drought used to get me down. I'd mope around and be discouraged. But as my faith in all areas has increased, I now take a completely different approach. When we're in a drought, I proclaim enthusiastically, "We're one day closer to rain, everybody!" It always rains, eventually. At this stage of my life, I've been through many drought cycles, and they always end with rain. This is not some sort of psychological gimmick; this is what watching faithful provision is all about. It's not just an academic Sunday School focus group. Provision is real, dramatic, and always on time. It might not be our timing, but it's on time in the bigger scheme of things.

In an urban environment, you don't get touched this dramatically and routinely with exercised faith. I'm not suggesting that you can't grow your faith in the city; I'm only pointing out the advantages a homestead platform offers. By the same token, a homestead doesn't guarantee a proper faith walk. But could there be a reason the Bible begins and ends in a garden? Was the agrarian base for the Israelites on purpose, or did they just happen into shepherding as a culture? How many of Jesus' parables centered around agricultural themes? Nearly all of them.

Cities carry hustle and bustle as a signature. Homesteads carry quiet and stillness as a signature. "Be still and know that I am God" is easier in the stillness of nature than it is waiting for a stoplight to change on Madison Avenue. Does that mean a New Yorker can't pray and walk simultaneously? Of course not, but that environment floods your senses with distractions. A homestead, by its nature, offers many hours of still quietness.

Your senses are not bombarded with the achievements of people; they're bombarded with the achievements of creation. The methodical pace of tending a garden, moving cows, and stacking hay all bring a clearer understanding of the systematic, ordered pace of life. Cities never sleep. Sirens, lights, and activity punctuate your senses like a perpetual percussionist on a drum set. On the homestead, you see the sun go down. It gets dark. The stars come out. The morning sky, ablaze with red diffusing through thin clouds, hales a new day. It can't be hurried. It's never late.

Season's cycles can't be hurried and can't be slowed. The final frost finally comes, and you can plant tomatoes. The first fall frost signals colored leaves, the end of peppers and tomatoes in the garden, and time to harvest sweet potatoes. Many faith groups put a huge emphasis on advent. A homestead screams, "This is the season for picking apples" at one time and at another, "This is the season for cutting firewood." I don't know anyone in the country who cuts firewood at night. But in the city, night and day blur together. Cycles blur together. The work stays the same. The routine stays the same. This monotony is why people want to get away.

Mentor to thousands, livestock management guru Bud Williams used to say that if you have to get away, don't come back. His point was that if you can't find fulfillment where you are, no place else will give it to you. You may as well relocate to a place or position that satisfies you if you can't be content where you are. By faith, you view your homestead as an unfinished canvas. One you can spend a lifetime working on without exhausting its nuances and capacities. Blossoming where we are, with what we've got, is an act of our will that we grasp by faith.

My mentor Allan Nation, founder of *The Stockman Grass Farmer* magazine, used to admonish me, "You have to learn to enjoy the slog." In other words, lots of life is dirty diapers and runny noses. By faith, you realize those chores are important. We move along each day knowing that faithfully applying compost will make the garden soil better. We select the best bunnies from a litter by faith that they will add vibrancy to our gene pool and the bunnies ten years hence will be better than the ones we have today. We dream and then dig to plant fruit trees, exercising faith that under our care, and with some sunlight and rain, they will produce a harvest in a couple of years. Maybe they'll feed our grandchildren.

Each day, as we wait for rain or sun, cold or warm, wind or still, we know we're part of something bigger than us. We step into a design and a plan that we accept by faith as fundamentally ordered, workable, and abundant. One of the most enjoyable aspects of working with animals is that they're predictable. Humans are emotional roller coasters. But my cows are always happy to see me. And they respond the same way to the same interactions. If I move too fast, they skitter away. If I lie down in the field, they come to sniffle me. If I call them, they follow.

Walking down an urban street approaching a stranger, you have no idea what that person is like. Is he friendly? Is he a villain? Will he rob me? Will he say hi? But you don't have that kind of variance with animals. Animals are as predictable as people aren't. I've never seen a cow not come to a bale of hay in the winter. Her needs are consistent and simple. With minimal demands, she satisfies easily. Some clean water, food, shelter if it's overly hot or cold. Cows don't see advertisements, so they don't know what they're missing. A cow never told me she wanted only hay mowed with a red machine instead of a green one.

After trying to please the boss, please customers, please co-workers, pleasing a cow is easy peasy. When all around is helter-skelter and topsy-turvy, we can retreat to the homestead where predictability bolsters our faith that the world isn't spinning out of control. We don't have to wonder if the cows will come to hay. Or if the chickens will scratch for worms. Or if the tomatoes will produce tomatoes. We can't see all the reasons why this true; we know that it is. That's a lot like faith, which is assuredness about things we can't see or completely understand.

How a seed sprouts is beyond our understanding. How a bull knows which cow is in heat and just the right hour to serve her is a lot of communication between the two we can't hear, see, or comprehend. Why a vine wants to send out curly tentacles to grab trellis wires and a pepper plant is content to stand upright without any support—these things are way beyond our comprehension. And yet they occur, every year, and assure us that order reigns. When everything seems like chaos, we understand that we can grasp the future by faith because of our interaction every day with the order of creation's design. We can know that in the end, the design works. God set it up that way.

Modern American culture seems fixated on chaos, on destroying nature's design and God's order. Our plants and animals know what they are and how to act. A pepper plant never sends out curly, grasping vine tentacles. A carrot doesn't mate with a radish. A pig doesn't wonder if it would be happier if it were a cow. Being this imbedded in a place that honors and respects the Creator's standards, boundaries, and designs brings solace to the soul. It massages our spirits with consistent common sense. It assures us that "God's got this." We need that.

Every day I have the privilege and honor to step outside the back door into a womb of provision, order, and abundance. I'm still in awe every day that the grass grows, the cows calve, the tomatoes ripen, and the grapes yield succulent juice. The magnificence and wonder of it all permeates my nose, ears, eyes, and touch. I get to be an extension of God's caress on creation, steward and caretaker of His benevolent object lesson of wonder and grace. A homestead represents all of that. I get to participate in that divine message. By faith I caress the homestead as hands and feet of the Creator.

If I made a world and put humans on it, if I made animals and plants and sunshine and rain, I'd have an idea about how it should be treated. I think I'd have a notion about what I'd like my people to put on it. May I be bold enough to suggest that

I'd like it adorned with homesteads. I wouldn't want it scabbed with factory farms, chemical-laden fields, and mega-processing facilities. I'd want it festooned with rainbows from functional evapotranspiration cycles.

Modern America is in a faith drought. More than half the people in our nation now consider themselves outside a religion. To my knowledge, no one has surveyed homesteaders on this issue versus urbanites, but my intuition is that country folks have a much higher faith response than city folks. Because the country puts us closer to the beauty, cycles, and conversations in nature, our ability to commune with something bigger than us develops easier. We're utterly dependent on seasons and soil, yes, but we're also awed daily by the majesty of living things. In many ways, homesteads bring us closer to faith. Our nation could use that.

Our children could use that. I need that. The more I can interact with the magnificence of things beyond human imagination, the more my view of myself and my place in an infinite plan aligns with faith principles. Homesteads foster a faith perspective, and that's good for all of us.

Chapter 20

The Wrap

Thank you for indulging me this exploration of what homesteads can do for our country, critters, and kids. In the Preface, I said I was writing this book to three people.

First, the person teetering on the precipice, wondering if taking the plunge to a homestead is worth it.

Second, the person who took the plunge a couple of years ago and is now in the slough of despond because the garden wilted, bugs ate the cabbage, and the neighbor's dog killed the chickens.

Third, the person who can't figure out why someone they love dearly and care about would head out to the country with snakes, dark nights, and nothing to do.

Let me take these one at a time with a final word of encouragement now that we've examined homesteading from many different angles.

TEETERING ON THE PRECIPICE

You're quite concerned about the future. You've read and heard the alternative voices, like trumpets on a wall, sounding the alarm about approaching changes. Just so we're all on the same page, let me itemize some of these things:

1. Increasing health problems from deficient food, side effects from orthodox medical treatments, and government regulatory intervention in what treatments are legal (like homeopathic remedies) and what insurance recognizes as acceptable.

2. Laws upon laws upon laws that move us toward tyranny, reducing freedoms and making all of us become criminals, even unintentionally and accidentally.

3. Frustration and concern about how to get a handle on our kids' increasingly addictive screen time; boundaries are harder to set, and parents feel like we're losing our kids down a hole of social hell.

4. A collapsing economy due to eroding money supply through profligate spending, Modern Monetary Theory, and complicated tax policy that makes financial survival more difficult.

5. Instability in the food supply, with prices, empty store shelves, and livestock diseases constantly featured as lead news stories.

6. Social breakdown in our cities, from crime to drugs to gangs.

7. The frenetic pace required to earn a living and keep up with friends' expectations regarding clothes, vacations, screen devices, recreation, and all the things that "keeping up" entails.

8. Wokeism, censorship, and increasing vitriol toward traditional Judeao-Christian beliefs, including an aggressive agenda toward sexualizing everything.

9. Feeling increasingly dependent on a system with a questionable agenda, from surveillance to tracking to Centralized Banking Digital Currency, and a yearning to simply disentangle.

10. The World Economic Forum agenda and other indicators tightening control on our lives, reducing freedom and choice.

I'm sure you have other items you could add to this list, but I kept it fairly generic on purpose; no need to irritate more than necessary. What you see is a drift toward places you don't want to go, and probably you've had some meltdowns as a result of a news story, asking, "What's going on in our world?"

People have left for better places throughout history. In fact, America was built by people who left someplace for a land that offered opportunity and freedom. Slavery is a blight on that history, but that changed to freedom too. Not perfect, to be sure, but on a steady journey toward a dream of opportunity. Although no paradise exists on this side of eternity, America experiments with opportunities hardly anyone in history could imagine.

You can make a list of things to be angry and frustrated about. Some will include my items above, and you can add others. We can either choose to wallow in that pit of disgust, worry and righteous indignation, or we can have our tantrum and then resolve to take all that negative emotional energy to a positive place.

That positive place is a refuge, an ark, a haven that offers hope and help to a society languishing in hopelessness and helplessness. We either can join the whiners or we can offer solutions. I hope by now you can see how a homestead offers solutions on many fronts. Don't worry about being the only one among your acquaintances who takes action. If nobody comes with you, that's okay.

This is not about who's NOT taking action; it's about those who DO take action. As we see breakdowns in numerous sectors of our society, and we mourn those losses, and even get angry about them, and fear for our future, we're hard-wired to flee. Our primal instincts kick in as we see goblins ahead and our natural protective inclination is to run away. That's natural and as it should be. But running away only lasts for a moment. At some point, you have to run toward something. Your fear must turn to faith. Your negative must turn positive. Only in running toward something better can you maintain your response action's staying power. As you make your moves, then, let fear be a catalyst but faith be the fuel.

Our responsibility is to the trumpets that sound in our soul, the alarm bells that awaken our spirit. How are we going to respond? The world is full of folks paralyzed between what they know they should do and those who actually do it. I have no illusions about this book creating a homestead stampede. That would be cool if it did, but I've lived long enough to know it won't. All that matters right now is you. What will you do?

BURNED OUT

Now I turn my attention to those of you who come up to me at homestead conferences, sometimes teary-eyed, feeling like a failure, looking like a fool to the friends who told you not to do it, and admit, "I don't know if we can do this. We can't get away, and it's so much work. Two goats died last week, and the three we still have got into the garden and ate all the tomatoes. We're so discouraged, and we're thinking this was a big mistake."

I hope as you've read the *why* you've found a bit of that first love. If it were easy, everyone would do it. But it's not easy, especially if you didn't grow up with any farming or homesteading in your background. This is a foreign country, foreign language, foreign everything. It takes time to assimilate. We didn't lose self-reliant understanding overnight and we won't find it overnight. Many of us are literally trying to undo generations of fleeing from the country; we won't rediscover what was lost in a year. Or even two. We're on a lifetime journey. My dad used to say the only thing worse than starting late is starting never. You started; that's half the battle.

First, dry your tears, get a glass of milk and a chocolate chip cookie and reconstruct the list you made when you decided to exit your urban life and become a homesteader. Why did you do it? See, you weren't crazy. You had really good reasons for making the change, for jumping into what you saw as a viable safety net. Unlike firefighters holding a safety net under the window of a burning building, though, your safety net wasn't put together yet. You have to put it together.

Once you're settled with your glass of raw milk and chocolate chip cookie, and your why list, look at your current situation and list your pain points. What are the things you feel most intimidated about? Now re-read *POLYFACE MICRO* to see if you fell into any of those "biggest mistakes." I'll bet you did. That's fine; most people do, and that's why I made the list. Take solace that you've gone the way most people go . . . at first.

Nobody gets it right at first. We don't talk well or walk well. Remember, if it's worth doing, it's worth doing poorly first. So you're doing it poorly. Big deal. Everybody goes through the toddler stage. Welcome to reality.

From my experience, let me offer a possible pain point list for you:

1. I can't keep the animals in; they get out and destroy things.
2. Chores take too long, and I don't have time to enjoy anything.
3. The kids complain because Papa John's pizza delivery won't come out this far.
4. We spent all that money on a Dexter cow, and she lost her calf.
5. The chainsaw is dull, and I don't know how to sharpen it.
6. Kids don't like to pull weeds.
7. A predator got in the laying hens and killed half of them.
8. All we do is carry buckets of water—to the livestock, to water the garden. Ugh!
9. The old farmhouse we thought was awesome is drafty and cold, and we freeze all winter.
10. We miss our city social life.

Did I get any right? Again, I've heard all these many times. You aren't the first one to go through this. It's almost like a drug addict going through withdrawal. The key is to substitute and not let yourself get into a vacuum. If you miss city social life, invite some neighbors over for dinner and establish new friendships. They won't invite you; you're new and weird. You be the one to reach out first.

Pull weeds with your kids. Turn it into a game. Read *FAMILY FRIENDLY FARMING*. The cure for pizza delivery is to make your own homemade pizza. You'll never buy pizza better than you can make at home; get everybody involved and turn it into a kitchen party.

Animal control is a proverbial Achilles heel with first-timers. Get some counsel from farmers in the area. Even conventional farmers can look at your situation and your animals and quickly advise on your weak points. Animals can be controlled; the problem is not them; it's your fences, energizer, or something you can fix.

You get the picture. I won't go through the whole list, but realize that every single pain point has a specific and doable solution. With dry eyes and clear thinking, brainstorm your solutions, prioritize them, and then begin attacking them one by one. Commit yourself anew to the why that brought you to this place; fall in love again with the opportunities; hug your family and resolve to prepare to win. Every problem has a solution.

One final thought, directly from my great friend Sina McCullough (co-author of *BEYOND LABELS* and co-host of our podcast by the same name): "Forgive yourself for not being able to meet all your expectations." That's a profoundly liberating thought. You came to this with all sorts of fantasy dreams, expecting it to be this way or that way. It didn't pan out exactly the way you thought. Sina uses this statement with parent-child issues, noting that she had a breakthrough as an adult when she came to the point that she could say to her parents: "I forgive you for not being what I wanted you to be."

We need to be able to say that to our homesteads. They do take on a life of their own, don't they? We view our homesteads almost as beings, as living creatures. Okay, forgive yours for not being all you wanted it to be. The world is an ocean; your homestead is a thimble. It can't hold or be or do everything you dreamed it would. It'll always take longer and be harder than you imagined. Welcome to anything that's worth doing; if it's easy, it's not worth much. Now dust off your britches and love your homestead again.

THE DUBIOUS

If you've read this because a friend asked you to, thank you. You've done one of the most courageous things a person can do, which is to expose yourself to some far-out ideas. Go ahead and congratulate yourself. How we respond to new ideas is the litmus test for how we view life. The greatest blessing of my life was growing up in a home that embraced being different. I buy and read books I know I'll disagree with just to expose myself to contrary ideas. Such exposure is good for your soul. When a new idea comes along, be the one to quickly say, "Tell me more." You don't have to agree, but be interested.

After trudging through this, if you still think your friend is nuts, throwing their life away or squandering their time and money, that's fine. It's a big world with plenty of room for different people. In fact, I'm really grateful for people like you because your patronage of my food keeps our farm in business. Everybody can't, won't, and shouldn't be a homesteader.

But I hope your friend's reasoning is clearer now, and while you may disagree, you can appreciate what's driving this yearning in her soul. And if you maintain friendship and interest in this grand adventure, perhaps she'll take you in and feed you when chaos comes. Or at least you'll be able to get something besides fake proteins saturated with antibiotics.

Don't rule out your own epiphany someday. Everyone comes to awareness at different times and due to different circumstances. Yours will undoubtedly come at some point too, hopefully before the door of the ark closes. Every person

who embraces doing something different has a testimonial
story about the conversion from sitting to standing to moving.
We even see it in the customers at our farm. Only a handful
grew up on nonchemical food. Most encountered a health issue
themselves or someone they loved and created an awareness
that moved them to a new way of thinking. Sometimes it's an
ecological disaster. It is seldom raw data; we are emotional
beings that normally don't change until touched emotionally.

You may never have to face difficulty. You may never
have something challenge your perceptions. I could wish
such an easy life on everyone. But that's not the way to bet.
Chances are you'll encounter things, like raindrops on your
head, that make you realize it's raining when others say it's
not. It all starts with a question. Usually, the question dares to
address the orthodox narrative. Could it be different than what
most folks say? What most news sources say? When people
question, make a change, and then find satisfaction in that new
direction, they almost always thirst for the next question and the
next change. Homeschoolers start gardening. Gardeners start
cooking. Cooks seek better ingredients. It's a progression. You
may not understand all the ins and outs of your friends' change
to homesteading, but you can be an encourager to the seeker.
That might be enough.

On our farm, which is a glorified homestead, we have
water, food, shelter, lumber, and winter heat. We know how to
build, grow, and repair. And we have enough people proximate
to pool together and at least be the last man standing. Nobody
can guarantee survival, but being the last man standing has its
own comfort because you hope, by that time, smarter minds
will have figured out how to solve the final problem. I don't
say any of this as a braggart; I say it as a grateful, blessed

player in a multi-generational desire to be less dependent on the conventional system.

When shocks come along and the headlines scream hysteria, our farm scarcely bobbles and we keep on going. The cows still eat their grass. The tomatoes still turn red. The chickens still lay eggs. The ponds still hold water. Whether the trend line is toward gradual dysfunction or is interrupted by a Black Swan event, biological and relational equity continues to function. We can take great solace in that as we love, nourish, and steward our homesteads.

I'm not a prophet, but I do agree with the Chinese proverb that says if we keep going the way we're going, we'll end up where we're headed. Many of us believe our civilization is headed in numerous unwise directions. We're raising a generation bereft of moorings and set adrift on an ocean of unwise and imprudent perceptions and expectations. For those who want to offer our civilization beacons of hope, arks of refuge, harbors of haven, homesteading is perhaps the most practical partner.

Whether you're thinking about it, tired of it, or baffled by it, homesteading is far more than backyard chickens and some homegrown tomatoes. Homesteads are like baptismal benchmarks signifying conversion from system entanglement to active participatory system disentanglement. The rewards are worth the effort. Thank you for joining the homestead tsunami for the good of country, critters, and kids.

Gonna Homestead Instead

By Joel Salatin

Traffic lights
Street fights
Siren nights
Gonna homestead instead.
Plant my garden
It's a bargain
Good for me
And family.

Traffic stops
Too many cops
Spinning like tops
Gonna homestead instead.
Settle life down
Outside of town
Good for me
And family.

Too much drugs
Lots of thugs
Sure need hugs
Gonna homestead instead.
Country friends
Make amends.
Good for me
And family.

Where's the food?
This ain't good
Here in the hood
Gonna homestead instead.
Plant my herbs
Outside burbs
Good for me
And family.

Don't see stars
Too many cars
Who needs bars?
Gonna homestead instead.
Make my meals
Wow this feels
Good for me
And family.

Too much noise
Artificial toys
Hard to find joy
Gonna homestead instead.
Out on the land
Birds for band
Good for me
And family.

Too much junk
Misfits and punks
Going into funk
Gonna homestead instead.
Room to run
That's more fun
Good for me
And family.

Work all day
Just for pay
Have no say
Gonna homestead instead.
Turn a new leaf
Grow my own beef
Good for me
And family.

Too fast pace
Leave no trace
Need more grace
Gonna homestead instead.
Scratch my pig
Plant and dig
Good for me
And family.

Watchin' TV
This ain't free
Binds me, see
Gonna homestead instead.
Working my hands
On my own lands
Good for me
And family.

INDEX

Other Books by Joel Salatin

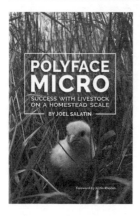

Polyface Micro: Success with Livestock on a Homestead Scale

Success with domestic livestock does not require large land bases. Polyface Farm leads the world in animal-friendly and ecologically authentic, commercial, pasture-based livestock production. In *Polyface Micro* Joel adapts the ideas and protocols to small holdings. Homesteaders can increase production, enjoy healthy animals, and create aesthetically and aromatically pleasant livestock systems.

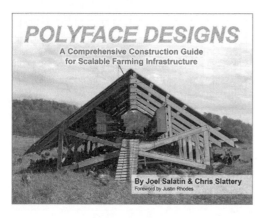

Polyface Designs: A Comprehensive Construction Guide for Scalable Farming Infrastructure

A comprehensive how-to manual of Polyface Farm's signature designs-with tips, tricks, and a half century of lessons learned through trial and error. Joel wrote the text and Polyface former apprentice and engineer extraordinaire Chris Slattery created the diagrams. Full color and beautiful enough to be a coffee table book even though you'll use it in your shop.

Beyond Labels: A Doctor and a Farmer Conquer Food Confusion One Bite at a Time

Joel Salatin and Sina McCullough bring you on a journey from generally unhealthy food and farming to an ultimately healing place. This book is designed to meet you where you are and motivate you to take the next step in your healing journey – ultimately bringing you closer to health, happiness, and freedom.

Your Successful Farm Business: Production, Profit, Pleasure

The sequel to Joel's *You Can Farm* builds on another 20 years of experience as Polyface Farm progressed from a small family operation to a 20-person, 6,000-customer, 50-restaurant business, all without sales targets, government grants, or an off-farm nest egg. Salatin offers a pathway to success, with production, profit, and pleasure thrown in for good measure.

The Marvelous Pigness of Pigs: Nurturing and Caring for All God's Creation

Growing up straddling the tension between the environmental and faith-based community, Joel pokes good-naturedly at the stereotypes with his self-acclaimed moniker: Christian libertarian environmentalist capitalist lunatic farmer. The question is simple: Do the beliefs in the pew align with what's on the menu?

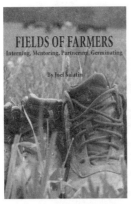

Fields of Farmers: Interning, Mentoring, Partnering, Germinating

America's average farmer is sixty years old. Our culture desperately needs a generational transfer of millions of farm acres facing abandonment, development, or amalgamation into ever-larger holdings. Based on his decades of experience with interns and multigenerational partnerships, Joel digs deep into the problems and solutions surrounding this land and knowledge-transfer crisis. This book empowers aspiring young farmers, midlife farmers, and nonfarming landlords to build regenerative, profitable agricultural enterprises.

Other Books by Joel Salatin

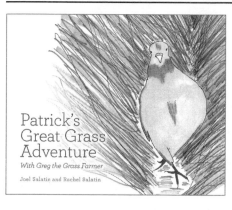

Patrick's Great Grass Adventure: With Greg the Grass Farmer

In his first children's book, Joel and his daughter Rachel Salatin team up on a whimsical tale about a pigeon, a farmer, and grass. Beautifully illustrated it introduces 4-7 year-olds to Greg the grass farmer through the eyes of Patrick Pigeon. What better way to discover ecology-enhancing grass farming than from an aerial view? Discover a real farm from a real farmer through captivating explanation and illustration.

Folks, This Ain't Normal: A Farmer's Advice for Happier Hens, Healthier People, and a Better World

Joel discusses how far removed we are from the simple, sustainable joy that comes from living close to the land and the people we love. Salatin understands what food should be: Wholesome, seasonal, raised naturally, procured locally, prepared lovingly, and eaten with a profound reverence for the circle of life.

Holy Cows and Hog Heaven: The Food Buyer's Guide to Farm Friendly Food

Written to empower food buyers in their dedication to food with integrity. Farmers who give it to their customers say that folks who have read it have a new level of understanding and a delightful attitude about the farmer-consumer relationship. Insights and real-life stories shared from Joel's own marketing experience.

Other Books by Joel Salatin

You Can Farm: The Entrepreneur's Guide to Start and $ucceed in a Farming Enterprise

Joel pulls from his eclectic sphere of knowledge, combines it with a half century of farming experience, and covers as many topics as he can think of that will affect the success of a farming venture. He offers his 10 best picks for profitable ventures, and the 10 worst. If this book scares you off, it will be the best reality check you ever bought.

$alad Bar Beef

Fishing for a phrase to describe this ultimately land-healing and nutrition-escalating production model, Joel coined the phrase Salad Bar to describe the farm's beef. Learn about herd effect, mobbing, moving, field design, water systems, manure monitoring, soil fertility, and even pigaerating. A fundamentally fresh way to look at the symbiosis between farmer, field, and cow. A classic in the pasture-based livestock movement.

The Sheer Ecstasy of Being a Lunatic Farmer

Can there really be that much difference between the way two farmers operate? After all, a cow is a cow and the land is the land, isn't it? Gleaning stories from his fifty years as localized, compost-fertilized, pasture-based, beyond organic farmer, Joel explores the differences. From how farmers view soil and water, to how they build fences, market their products or involve their families, this book shows depth of thought. Salatin explains a different food model and shows with good humor and stories how this alleged lunacy actually offers a life of sheer ecstasy.

Pastured Poultry Profit$: Net $25,000 on 20 Acres in 6 Months

Joel began raising chickens when he was 10 years old and fell into the pastured poultry concept a couple of years later. Still the engine that drives sales, notoriety, and profit, pastured poultry has revolutionized countless farming endeavors around the world. A how-to book, it includes all the stories and tips, from brooding to marketing. Centered around meat chickens, it includes a section on layers and turkeys. This book started the American pastured poultry movement.

Everything I Want to Do Is Illegal: War Stories from the Local Food Front

Although Polyface Farm has been glowingly featured in national print and video media, it would not exist if the USDA and the Virginia Department of Agriculture and Consumer Services had their way. From a lifetime of noncompliance, frustration, humor, and passion come the behind-the-scenes real stories that brought this family farm into the forefront of the non-industrial food system.

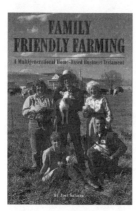

Family Friendly Farming: A Multi-Generational Home-Based Business Testament

Few life circumstances are as hard to navigate as family business. This book describes the rules and relational principles to harmonize in what is too often a tense environment. The chapters on how to get your children to enjoy working with you are worth the price of the book. Beyond that, it delves into the quagmire of inheritance, family meetings, and personal responsibility. A pathway exists to leverage the strengths of family business and hold families, and especially family farms, together.